Science

YEAR 4

Sue Hunter
Jenny Macdonald

GALORE PARK

AN HACHETTE UK COMPANY

Every effort has been made to trace all copyright holders, but if any have been inadvertently overlooked, the Publishers will be pleased to make the necessary arrangements at the first opportunity.

Although every effort has been made to ensure that website addresses are correct at time of going to press, Galore Park cannot be held responsible for the content of any website mentioned in this book. It is sometimes possible to find a relocated web page by typing in the address of the home page for a website in the URL window of your browser.

Hachette UK's policy is to use papers that are natural, renewable and recyclable products and made from wood grown in well-managed forests and other controlled sources. The logging and manufacturing processes are expected to conform to the environmental regulations of the country of origin.

Orders: **Teachers** please contact Hachette UK Distribution, Hely Hutchinson Centre, Milton Road, Didcot, Oxfordshire OX11 7HH. Telephone: (44) 01235 400555. Email: primary@hachette.co.uk. Lines are open from 9 a.m. to 5 p.m., Monday to Friday.

Parents, Tutors please call: (44) 02031 226405 (Monday to Friday, 9:30 a.m. to 4.30 p.m.). Email: parentenquiries@galorepark.co.uk

Visit our website at www.galorepark.co.uk for details of other revision guides for Common Entrance, examination papers and Galore Park publications.

ISBN: 978 1 471856 31 0

© Hodder & Stoughton 2015

First published in 2015 by Hodder & Stoughton Limited
An Hachette UK Company
Carmelite House
50 Victoria Embankment
London EC4Y 0DZ

Impression number   10 9 8

Year   2023

Cover photo © guidenop/iStock.com
Illustrations by Aptara, Inc.
Typeset in India by Aptara, Inc.
Printed and bound by CPI Group (UK) Ltd, Croydon, CR0 4YY

A catalogue record for this title is available from the British Library.

About the authors

Sue Hunter has very recently retired as full-time science teacher but she continues to be very involved in Science education. She is a member of the Common Entrance setting team and a governor of local primary schools. Sue has written extensively, including Galore Park's KS2 Science textbooks and 11+ Science Revision Guide.

Jenny Macdonald has had a happy and fulfilling career as a teacher, teaching in both state and private schools. She has recently retired having spent the last eighteen years teaching science in a local prep school. In the last few years she has contributed to Galore Park's KS2 Science textbooks.

Photo credits

Contents

Introduction

➔ About this book

To the teacher

Equipped with his five senses, man explores the universe around him and calls the adventure Science. **Edwin Powell Hubble**

The study of science for young children is a voyage of discovery. It stimulates their curiosity and provides a vehicle for them to explore their world, to ask questions about things that they observe and to make sense of their observations. It does not exist in isolation but draws upon many other aspects of a well-rounded curriculum and should be practical, interesting and, above all, fun.

This book covers the requirements for the National Curriculum for Year 4. It also contains additional material as necessary to meet the specification for Year 4 in the ISEB Common Entrance syllabus and some extension material. It includes ideas for activities to develop practical skills, deepen understanding and provide stimulus for discussion and questioning.

Practical work is always popular, and hands-on activities in this book are designed to be carried out by the pupils in pairs or small groups. Pupils should be encouraged to think about safety at all times when carrying out practical activities. However, the responsibility for risk assessment lies with the teacher who should ideally try out each activity before presenting it to the class in order to identify any risks as appropriate to the particular group of children involved. The ASE publication *Be safe!* (available via the ASE website: www. ase.org.uk/resources) is a useful source of information and advice about risk assessment in the primary phase.

Exercises have been set at intervals throughout the book. Where there is more than one exercise in a group, the first one is set at standard level, followed by a more easily accessible exercise covering the same material and/or an extension exercise.

To the student

This book is to guide you in your study of science in Year 4. Science is a fascinating subject because it tells you so much about yourself and the world you live in. Science is all about asking questions and finding answers to them, so use the information in this book as a starter but remember to look, listen and ask questions to take you further.

There are some special features in the book that are designed to help you in your work.

⊙ Notes on features

Words printed in **blue and bold** are keywords. All keywords are defined in the glossary towards the end of the book.

Exercise

Exercises of varying lengths are provided to give you plenty of opportunities to practise what you have learned.

Activity

Sometimes it is useful to explore a topic in more detail by researching it. An activity is an opportunity to discover interesting things for yourself, and to practise recording and presenting what you find out. Some activities provide opportunities for you to do experiments. Others need some research from books or the internet, or maybe by talking to other people.

Did you know?

In these boxes you will learn interesting and often surprising facts about the natural world to inform your understanding of each topic.

Working Scientifically is an important part of learning science. When you see this mark you will be practising the really important skills that make good scientists. You will find out:

- why we carry out experiments

- what we mean by the word 'variable'

- what we mean by a fair test

- how to design experiments to answer your own questions

- how to measure variables

- how to record and display results clearly and accurately

- how scientific understanding is built up by the work of many scientists learning from each other, sometimes over hundreds of years.

 Go further

The material in these boxes goes beyond the ISEB syllabus for 11+. You do not need to learn it for an 11+ exam but your teacher may decide that it is a good idea for you to learn something a bit extra to help you to understand a topic better or to extend your learning. All this material will be useful to you in your future studies ...

The variety of living things

The Earth is a huge place. Across the surface of the Earth there are many different places with different conditions. Some are hot, some are cold. They may be dry or wet, high or low, wild or inhabited by humans. Each of these different places will be home to a different range of plants and animals.

There are thousands of different types of animal and plant, many of which have not yet been discovered. Sadly, many types of living thing may become extinct before we even know they exist. However, scientists exploring remote places still find new animals and plants to study.

➲ Making groups

It is easier to learn more about the animals and plants in a particular place if we sort them into groups. To do this we need to find particular characteristics that they have in common. There are many different ways we could do this and we have to choose sorting characteristics that put them into useful groups.

For instance, we might choose to sort animals by colour. The green group might include a grasshopper, a snake, a frog and a parrot. This is not a very useful group – apart from their colour, these animals do not have much in common.

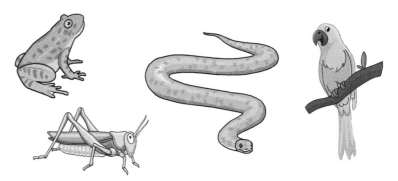

These green animals do not have much in common with each other

Scientists have developed a system that does make useful groups. You will learn more about this system in Year 6 but we can start by looking at some of the groups and which animals and plants belong in them.

➔ Vertebrates

In Year 3 you learnt that some animals have bony skeletons inside their bodies. We group these animals together and we call them **vertebrates**.

This is still quite a big group so we usually divide it up into five smaller groups: **mammals**, **birds**, **fish**, **amphibians** and **reptiles**.

Mammals

Mammals are animals that have fur on their bodies. Female mammals have special places in their bodies where they make milk to feed their babies. You are a mammal and so are dogs, cats, horses, elephants and shrews. Any animal with a furry body is a mammal so they are quite easy to spot.

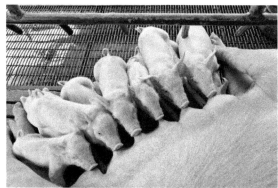
The mother pig feeds her babies with milk. Can you see the fur on their bodies?

The fur on the bodies of mammals is very good at keeping them warm in cold places and cool in hot ones. This means that mammals can live in lots of different places.

Did you know?

Whales and dolphins are mammals, even though they have almost no hair on their bodies. They have lost the hair on their bodies to make them more streamlined as they race through the water. Instead of fur, they have a very thick layer of fat called blubber under their skin to keep them warm in the cold seawater. Changes like these, that make an animal or plant especially well suited to living in a particular place, are called **adaptations**. You will learn more about adaptations in Year 5.

1   The variety of living things

Birds

Birds are animals that have feathers on their bodies and their front limbs are wings. Their babies form inside hard-shelled eggs. Examples of birds include blackbirds, eagles and ostriches.

Most birds use their wings to fly and the feathers help them to do this. Feathers also help to keep them warm.

 Did you know?

Scientists believe that birds developed from small dinosaurs. Some dinosaur fossils have been found that have signs of feathers. A very special fossil called Archaeopteryx was found in Germany in 1861. It is a kind of dinosaur but looks very much like a bird.

Did you know?

Birds can be all different sizes. The smallest bird is a type of humming bird, which is not much bigger than a bumblebee. The largest bird is the ostrich, which is so big that it could not possibly fly.

■ Ostriches are too big to fly

Fish

All fish live in water. Some live in the fresh water of lakes and streams and some live in salty seawater. They have scaly bodies and can breathe underwater using gills.

Fish do not have any legs or arms. Their bodies are smooth and streamlined to help them to swim through the water. Their tails are shaped like paddles to help them move forward and their fins help them to steer.

Did you know?

The mudskipper is a type of fish that spends some of its time out of water. It can use its front fins a bit like legs, so it can 'walk' across the mud. It is able to hold a supply of water around its gills while it is out of the water and it can also take in oxygen from the air through its skin.

■ The mudskipper is able to survive out of water for a short time

Amphibians

Amphibians are animals such as frogs, toads and newts. They spend some of their time in the water and the rest in damp places on land. Their babies, called **tadpoles**, look a bit like little tiny fish at first. You may have watched little frog tadpoles hatching from the jelly-like **frog spawn**. They soon change into tiny little froglets. You will learn more about this in Year 5.

■ Frogs are amphibians

Did you know?

Baby amphibians live in the water so they breathe through gills, like fish. Adult amphibians have lungs so they can breathe in the air. They can also breathe through their skin when they go back into the water. They can stay submerged for a very long time because their skin takes in oxygen from the water.

Reptiles

There are many different reptiles. Sometimes it is hard to see how they can all belong in the same group. Snakes, lizards, tortoises and crocodiles are all reptiles. They all look very different but they do have some things in common. They all have dry, scaly skin and they lay eggs with leathery shells.

Did you know?

Crocodiles are very ancient animals. They are still very much like their ancestors, which lived millions of years ago alongside the dinosaurs.

Tortoises are also quite ancient. They too have ancestors that lived in the time of the dinosaurs. Tortoises are also the longest living land animals. Some tortoises are known to have lived for over 150 years!

■ Tortoises can live to be over 150 years old

Exercise 1.1

Here is a list of animals. Think about each one carefully and then write down whether each is a mammal, a bird, a fish, an amphibian or a reptile. If you find it hard to identify any of them, do some research in a book or on the internet to help you.

1 mouse

2 tree frog

3 polar bear

4 boa constrictor

5 golden eagle

6 shark

7 turtle

8 natterjack toad

9 tuna

10 penguin

➲ Invertebrates

When you have been outside learning about the environment near your school, you have probably spent some time looking for small animals, often known as minibeasts. These animals do not have bony skeletons inside their bodies so their proper name is **invertebrates**. You learnt a little about these in Year 3. Some of them have soft bodies and some have hard cases called **exoskeletons** all around their bodies.

Nobody knows exactly how many different types of invertebrate there are. New ones are being found all the time and some of them are so small that you wouldn't notice them at all without a microscope. One estimate is that there are probably about 1.3 million different types. As with the vertebrates, it is useful to put them into smaller groups to help us to study them. As there are so many, we will not look at all the groups in this book, just the ones you are most likely to come across in your school grounds or garden.

Invertebrates with exoskeletons

The animals with exoskeletons all have legs. The exoskeleton makes the legs strong enough to hold the animal up and they have joints in their legs so they can usually run quite fast. There are more of these animals than any of the other types so it is quite helpful to sort them into smaller groups. We can do this by counting their legs.

Invertebrates with six legs

You may already know that animals with six legs are called **insects**. Many of these have wings and they all have bodies divided into three parts. The insects are the biggest group of animals on Earth. They include animals such as flies, wasps and beetles.

There are some very big insects, such as the Titan beetle and the Morpho butterfly. The smallest known flying insect is a type of wasp that is just 0.15 mm long. Millions of years ago there was a type of dragonfly that had a wingspan of 2.4 m!

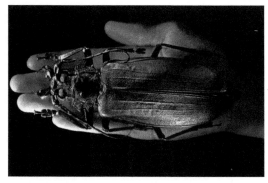

◼ The huge Titan beetle has jaws so strong that it can bite through a pencil!

Did you know?

Without insects we would be in serious trouble. Some insects carry pollen from one flower to another (you will have learnt about this in Year 3). If they did not carry out this important task, plants would not be able to make seeds and we would be without most of the different fruits and vegetables that we need for a balanced diet.

Invertebrates with eight legs

Animals with eight legs belong to the **spider** group. This includes all the different types of spider that you might see running around in your house or garden, from the big black hunting spiders to the little 'money spiders' and the brown house spiders you might see in an empty bath. It also includes scorpions and tiny eight-legged mites.

Spiders have two parts to their bodies. They are carnivores and some of them build beautiful webs to help them to catch insects.

■ The spider's web helps it to catch its insect prey

 Did you know?

Many spiders have a poisonous bite to help them to kill their prey. Some, like the black widow in Australia, can even be deadly to humans.

You will not be killed by any spider native to Britain but many of our spiders do have a poisonous bite. However, they have such weak fangs that they are unable to pierce our skin and their poison is not very strong.

→ Invertebrates with more than eight legs

As well as the two well-known groups above there are two other groups of invertebrates with exoskeletons, the **crustaceans** and the **myriapods**.

Crustaceans have ten legs. This group includes crabs, prawns and lobsters. Most of these live in water but some crabs live on land. The land crustacean that you are most likely to find is the woodlouse.

Myriapods are animals with more than ten legs. This includes centipedes and millipedes.

■ Crabs and lobsters have ten legs. The front pair have become pincers

Did you know?

What is the difference between a centipede and a millipede? Many people think that a centipede has a hundred legs and a millipede has a thousand. In fact they both have much fewer legs than this. Both groups have long, thin bodies divided into many sections.

Centipedes are ferocious hunters. They have one pair of legs on each body section and the legs stick out at the sides. They are often orange or brown in colour and can run very fast.

Millipedes feed on dead leaves and rotting wood. They are usually greyish brown in colour. They have two pairs of thin spindly legs on each body section and these are tucked away under their bodies, making them look a bit like worms.

Invertebrates without exoskeletons

This group includes worms, slugs and snails as well as some other animals, such as jellyfish and corals.

Worms, slugs and snails all have soft bodies with no internal or external support. There are several different types of worm and each is found in a different place. Earthworms have long, flexible bodies that allow them to wriggle their way through the soil. They eat dead leaves and often drag them down into their burrows where they can eat them safely. They are very important to us because they help to make the soil better for growing plants in (you will have learnt about this in Year 3). They are also an important part of the diet of many animals, such as blackbirds and hedgehogs.

Worms are an important part of the diet of many animals

Slugs and snails are very similar animals. Both of them slide along the ground on their tummies and their proper name (gastropod) means 'stomach foot'. They have **antennae** on their heads and these have simple eyes on the end. They cannot see very well though. Snails carry hard spiral shells on their backs and, although these must make moving around a bit more difficult, the snail can pull its soft body back into the shell if it is attacked, making it harder for a predator to eat it. Both slugs and snails eat leaves and will happily eat the lovely tender vegetables and other plants growing in our gardens.

The spiral shell of the snail helps to protect it from predators

Did you know?

If you collect the shells of snails and look at them carefully, you will see that they are coiled into a spiral. Some coil towards the snail's right-hand side and others coil in the opposite direction. The Roman snail occurs in the west and south of England. The shell of the Roman snail almost always coils to the right side of the snail's body. However, every now and then there might be a rare 'left-handed' one. Some snail experts call these 'snail kings'.

Exercise 1.2

Match each type of invertebrate in the left-hand column, to the correct description in the right-hand column.

insects   animals with an exoskeleton and four pairs of legs

spiders   animals with soft bodies that are long, thin and flexible

worms   animals with soft bodies that slither along on their stomachs

slugs and snails   animals with an exoskeleton and six legs

➲ Plants

Plants can also be divided into groups in various different ways. Scientists usually start by sorting them into two big groups. One group of plants produce flowers and seeds to reproduce and the other group does not produce flowers and seeds. We call these two groups **flowering plants** and **non-flowering plants**.

Flowering plants

You might think that it is always easy to spot a flowering plant. We grow many of them in our gardens and parks because the flowers are so bright and cheerful. We can see the flowers on the bushes in a hedgerow or on a big horse chestnut tree. You learnt in Year 3 that these bright colours attract insects to come to the flowers and pollinate them.

Some flowering plants have flowers that you may not notice at all. Have you ever spotted the flowers on an oak tree or a hazel bush? Did you know that the fluffy tops on long grass are flowers? These plants all have flowers and make seeds but they are not bright and colourful. They are not pollinated by insects so they do not need to make bright showy flowers. Instead their flowers hang down from the rest of the plant and

■ The red flowers of the poppies in the meadow are easier to spot than the fluffy grass flowers

wait for the wind to carry the pollen from one flower to another. If you suffer from hay fever in the summer, it is the pollen from wind-pollinated plants that causes your discomfort.

 Did you know?

Some flies like to lay their eggs in decaying animal bodies. The larvae then feed on the rotting flesh before turning into adult flies. Some plants trick these flies by making a smell like rotting meat. The flies go into the smelly plant before they discover their mistake, get covered in pollen and then carry it away and pollinate the next plant that tricks them.

Non-flowering plants

Mosses and ferns are examples of non-flowering plants. This group also includes seaweeds and the algae that make green patches on the trunks of trees and in other damp places. These plants do not make seeds so they do not need flowers.

 Did you know?

If you look at the roof of your school or house, you may see some rounded clumps of moss, called moss cushions. Each of these moss cushions is a tiny world of its own. If you squeeze the water out of a moss cushion and look at it through a microscope you will probably see several different types of tiny invertebrates swimming around. This might include a little animal called a tardigrade (or water bear).

A tardigrade (water bear) is less than 1 mm long

Use the following words to help you to fill in the gaps in these sentences. Some words may need to be used more than once.

**amphibians   exoskeleton   five   flowering   groups   insects
invertebrates   mammals   millions   mosses   predators   reptiles
skeleton   slugs   snails   spiders   study   vertebrates   wind**

1 There are _____ of different types of plant and animal on Earth. Scientists sort them into _____ to make it easier to _____ them.

2 Animals are divided into two big groups. Animals with an internal _____ made up of bones are called _____. Animals without bones are called _____.

3 The animals with skeletons are divided into _____ smaller groups. These are _____, birds, fish, _____ and _____.

4 Some animals without bones have a hard outer case called an _____. These include _____, which have six legs and _____, which have eight legs.

5 Worms, _____ and _____ have soft bodies. _____ carry a hard shell on their backs for protection from _____.

6 Plants can be divided into two groups. _____ plants may have brightly coloured petals to attract _____ but many have small, dangling flowers, which are pollinated by the _____. Non-flowering plants include _____ and ferns.

Rebecca found some animals in her garden. She wrote some clues about them and then challenged her friends to identify them.

1 See if you can work out what animals she found.

(a) This animal has six legs and a hard exoskeleton. It is rounded in shape and has a red back with black spots.

(b) This animal has four legs and a furry body. It has a long tail and is very good at catching birds and other small animals.

(c) This animal has no legs. It has a soft body and antennae on its head. It slithers across the ground and does not have a shell.

(d) This animal has four pairs of legs. It has a tiny body and very long, thin legs.

(e) I found this animal in the soil. It has a long, thin, soft body.

2 Now say which animal group each of Rebecca's animals belongs to.

3 You could make up some clues like Rebecca's to try on your friends. Remember to ask them to say which animal group each one belongs to.

Finding the name

When we find an animal or plant we sometimes know its name but often we do not. There may be someone nearby who can tell you the name but there are so many different types of animal and plant, even in our local area, that most people will not be able to identify them all. When this happens we need something to help us. Sometimes we can find the answer in a book called a field guide. These books have pictures and information about living things. They divide the living things up into groups, as we have in this chapter, to make it easier to find them. Sometimes it can take us quite a lot of searching to find the right name even using a field guide. We need some more clues to make it easier.

One way to do this is to use a classification **key**. There are two types of classification key: branching keys and number keys. To use a key you need to be observant and to notice the differences between living things.

Using a branching key

A branching key looks a bit like the roots of a tree. At each junction there is a question. Each question can be answered in one of two ways, usually 'Yes' or 'No'. By answering all the questions you can work your way through it to find the answer.

Here is a simple one for you to try.

Look at these pictures of British butterflies. Then use the key below to find the name of each one.

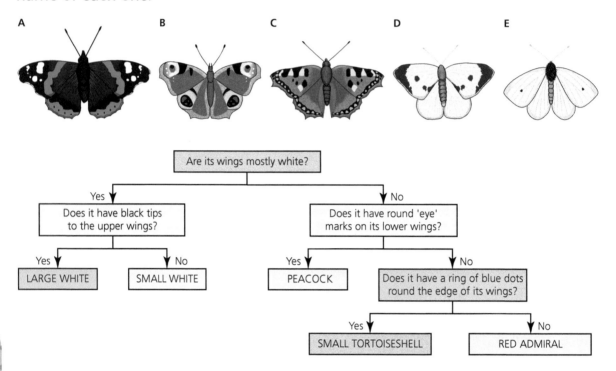

Using a number key

A number key has a series of numbered clues for you to work through. To use a number key, you start at clue number 1 where you will be given two possible alternative descriptions. You need to decide which of the two matches the thing you are trying to identify. At the end of the line with this description, you will find an instruction telling you which clue to go to next. This clue will give you two more choices and so on until you have identified your item.

Here is a simple example for you to try.

Look at the drawings of some birds you might find in a woodland.

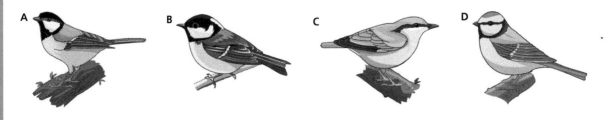

1	The bird has a black head.	Go to Clue 2
	The bird does not have a black head.	Go to Clue 3
2	The bird has green feathers on its back.	GREAT TIT
	The bird does not have green back feathers.	COAL TIT
3	The bird has yellow feathers on its breast.	BLUE TIT
	The bird does not have yellow feathers on its breast.	NUTHATCH

Most keys are longer than these ones and allow you to identify more living things. Some keys are quite complicated, but they all work in one of these two ways. Each is either a branching key with questions or a number key with clues. If you go out to do some fieldwork with your teacher, you will probably have some keys with you to help you to identify the animals and plants you find.

Exercise 1.4

1 Use the number key below to identify which trees the following leaves came from.

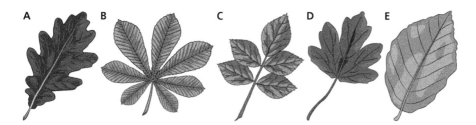

A B C D E

1	The leaf is made up from smaller leaflets.	Go to 2
	The leaf is one solid shape.	Go to 3
2	The leaflets fan out from a central point.	HORSE CHESTNUT
	The leaflets are arranged along the stem.	ELDER
3	The leaf has a smooth oval shape.	BEECH
	The leaf has wavy shapes around the edges.	Go to 4
4	The leaf veins fan out from the top of the stalk.	FIELD MAPLE
	The leaf has a central branching vein.	OAK

2 Use the branching key below to help you to identify these toadstools.

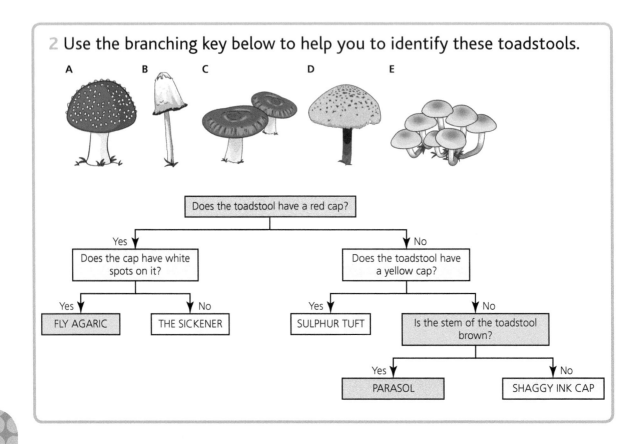

Activity – field study

Spend some time just looking and listening in a local place. This could be the school grounds, your garden, a local woodland or park, by a pond or at the seaside. Your teacher may give you some keys or a field guide to help you to identify the living things you find.

Sit very quietly for a while and think about what you can hear and see. Try to identify any birds or other larger animals that you observe. Remember to include dogs, horses, farm animals, humans, etc. Write them all down in your notebook. If you are unable to identify an animal, you could draw a quick, labelled sketch in your book. You can then come back to this later to identify it.

Now think small. How many invertebrates can you spot? You may need to turn over some stones or logs to look under them but remember to put them back in the same place immediately when you have finished looking. Look on the undersides of leaves and in cracks and crevices in the bark of trees or in walls – anywhere where an invertebrate might hide. You do not need to catch the animals. You just need to see and identify them. Add your invertebrates to the list, with a labelled sketch if you are unable to find its name.

Now think about the animals you cannot see but which use the area. You might see signs of them, such as footprints or the remains of a meal. Perhaps you can see a burrow or nest. Add these invisible animals to your list.

Now think about the plants. Which ones can you name? Can you find out the names of others? Add these to the list. Remember to think about the less noticeable non-flowering plants, such as mosses and algae. Make a list of the plants you have spotted.

When you get back to the classroom, discuss your list with your partner or group. Can you sort all the animals and plants into the correct groups? Use this chapter to help you.

2 Caring for the environment

The word **environment** means the surroundings. It is often used to mean the whole of the natural world; in other words the surroundings of every living thing on the planet.

➲ Living things in their habitats

Each type of living thing is found in a particular part of the environment. Different living things live in places with different conditions and we call the special place where an animal or plant lives its **habitat**. A habitat has to provide everything that an animal or plant needs to survive.

In Year 3 you learnt about the life processes that all living things carry out to survive. How many of them can you remember? A good habitat is a place where living things can carry out all the life processes.

All animals need a supply of water and the right kind of food for the life process we call **nutrition**. Animals need to be able to move freely around the habitat to find food and escape predators (movement life process). They need safe places to build a den or nest where their young can be born and grow up (**reproduction** and growth life processes).

Plants need light, water and minerals from the soil for nutrition. They need insects or the wind to pollinate their

■ A nest box in a woodland provides a safe place for this great tit to rear a family

flowers and somewhere suitable for their seeds to grow for reproduction. All living things also need a supply of oxygen from the air or water for **respiration**.

Look at this picture of a woodland habitat.

There are many different types of plant and animal that live in this habitat. The woodland contains several different types of tree. There are lots of smaller plants as well. Many insects, spiders and other invertebrates live in the trees or in the leaf litter on the ground. Hedgehogs, badgers, squirrels, mice and deer are all found in the woodland. Robins, blue tits, woodpeckers and other birds also make this habitat their home. The woodland provides them with shelter, food and somewhere safe to reproduce.

A woodland habitat provides many places for animals and plants to live

Activity – life processes in the woodland

One animal that lives in the woodland is a squirrel. Here is how the squirrel carries out the most important life processes in the woodland.

Life process	How the squirrel carries out the life process
Nutrition	It eats acorns, hazelnuts and other seeds. It also likes to eat fruit such as blackberries. It stores nuts and seeds in the autumn to eat during the winter.
Movement	It runs about on four legs, using its tail for balance. It can move around on the ground but is also very good at running around in the trees and jumping from branch to branch.
Reproduction and growth	It builds a ball-shaped nest out of twigs, grass and moss in a tall tree. The young are born in spring and grow quickly. The mother feeds them on milk until they are old enough to eat nuts and seeds.

Work with your class to find out more about how other animals and plants in the woodland, or another habitat that you have studied, carry out these life processes. Choose one or more living things that might be found in the habitat but try to share them out so that you find out about as many different living things as possible. Remember to think about the invertebrates and the plants as well as the larger animals.

Use books and the internet to find out more about your chosen living things and, for each one, make a table like the one on the previous page to show how the animal or plant carries out the life processes. Make a neat, clear copy of the table and then try to find a good picture of each of your living things. Put all the research together into a big scrapbook or onto a wall display so that other people can find out about the animals and plants in this habitat.

The woodland in the picture on the previous page is just one of many different habitats that you might find near where you live. Animals and plants live in parks and gardens, in roadside verges and trees. You will find living things in ponds and lakes and on the seashore. Farmland and moorland are also habitats for some animals and plants.

■ Roadside verges are good habitats for many plants and animals

In each of these habitats animals and plants need to be able to carry out the life processes. Sometimes the way that they do this has to change because of the special conditions in the habitat. For example, a river is a very different place to a woodland. Fish live in the water so they need to be able to breathe oxygen from the water rather than the air. They need to find ways of reproducing without their eggs or young being swept away by the moving water. These special differences in the way that animals survive in different habitats are called adaptations. You will learn more about adaptations in Year 5.

Exercise 2.1

1 What is meant by the term 'environment'?

2 What name do we give to the special place where an animal or plant lives?

3 What must this special place provide for the animals and plants that live there?

4 Name three life processes carried out by all living things.

⊙ The naturally changing environment

Animals and plants need to be able to cope with changing conditions in their habitats. What changes do you think the animals and plants in the woodland might experience? Will the conditions in the woodland be the same all year round?

■ The woodland looks very different in the winter

In Britain and other similar countries, habitats change a bit with the **seasons**. The weather in summer is warmer than in winter. Many trees lose their leaves in the winter and grow new ones in the summer. Some of these changes can make life harder for the animals and plants that live there. For example, strong winds may break branches off trees or even blow them over. This is hard for the tree. It is also hard for the animals that rely on the tree for food or shelter.

In some places the changes are much more difficult for the living things to cope with. For example, in parts of Africa there are just two seasons: the rainy season and the dry season. In the dry season there is no rain at all and it may become so hot that all the rivers dry up and all the plants shrivel and lose their leaves or die.

Any change to the environment is hard for living things. They have to change the way that they live. They may find it harder to find enough food or somewhere to breed. Changes happen naturally in environments. This may be because of the changing seasons, or it could be because of a natural disaster, such as a flood or a volcano erupting.

■ The dry season in Africa can be very hard for the animals that live there

⊙ Humans cause changes to the environment

There are many ways in which the environment changes naturally. However, there are also many ways in which humans can cause changes to the natural world. If you look out of the window, you will see a landscape that has been changed by humans in some way.

■ How have humans changed this landscape?

Activity – how have humans changed the environment?

Look carefully at the environment around you. You may be looking out of a window, standing in the school grounds or maybe your teacher has taken you to a local park or woodland.

Think carefully about what the place you are looking at might be like if there were no humans. Discuss with your partner or group all the ways in which you think that humans might have changed the environment where you are. Compare your ideas with the rest of the class. Do you think the changes will have affected the wildlife? Do you think the changes have made the environment better or worse for the living things that live there, including humans?

All animals affect their environment in some ways, but humans have much more effect than other animals. There are billions of humans and each one needs food, water and shelter. We have taken over large areas of the natural world to grow food. We use up resources to make our homes and other buildings. We build huge towns and cities where once there were woods and prairies. We take lots of water from rivers and lakes.

Humans like to travel and people want lots of things to make their lives easier and more comfortable. Humans have built factories, cars, lorries and aeroplanes, which give out gases that **pollute** the air. We also dig up coal, oil and other resources, causing damage to habitats and disturbing the wildlife in them.

How does all this human activity affect the animals and plants that share the planet with us? The answer is that what we do affects them a great deal. If we understand how we are harming wildlife we can begin to take steps to solve the problems we create. We can think about how we can change the way we live so we do less damage, and how to put right any damage that we have already done.

Pollution

Many human activities put things into the environment that shouldn't be there. We call this **pollution**. Let's think about some ways in which our actions can cause pollution and what we might do about it.

Human activities make a lot of rubbish. If we do not dispose of our rubbish in a responsible way it can cause problems for wildlife. Even dropping a small piece of litter can be harmful. Animals may be hurt or they may try to eat something that is not good for them. Rubbish can cover up plants and kill them and it doesn't look very good in the environment either. The picture shows some rubbish that has been dumped by the roadside.

What can we do to make sure that our rubbish causes as little harm to the environment as possible? Discuss your ideas with your partner or group.

Sometimes waste materials from factories spill out into rivers and lakes. When farmers spray their fields with chemicals, some of the chemicals may wash off into the river when it rains. The chemicals in this river have killed all the wildlife.

What could people do to stop this type of problem? Discuss your ideas with your partner or group.

We use a lot of energy. Our electricity is made in power stations that burn coal, oil or gas. Our cars burn petrol or diesel, which are made from oil. When we burn coal, oil or gas we put harmful gases into the air. There are lots of effects that these can have.

Some gases are acids and cause **acid rain**, which kills plants and may poison the water in lakes and ponds. Some gases get stuck in the blanket of gas that surrounds the Earth, called the **atmosphere**. These stop heat from escaping into space, and as we add more of them to the atmosphere, the Earth is slowly getting warmer. This causes problems for many animals and plants.

The picture shows a thick, choking fog, often called smog. This is caused by gases and dust from the coal burnt in the power stations nearby. It makes people ill as well as harming plants and animals.

How do you think we can help to reduce the amount of polluting gases added to the air? Discuss your ideas with your partner or group.

Did you know?

The 2008 Olympic Games were held in Beijing in China. As the time of the games got closer, Beijing was covered in a thick yellowish smog, like the one in the photograph. To clear the air for the games, power stations and factories were closed down to stop them from adding more pollution to the air.

Activity – pollution

One common form of pollution is oil from ships being spilt into the sea. The oil is poisonous and it also sticks to any animals or plants that come into contact with it. Birds with oil on their feathers cannot fly and they cannot catch food. Cleaning up oil pollution is not easy.

You will need:

- two bowls

- water

- olive oil or vegetable cooking oil

- a feather

- a spoon

- some washing-up liquid and a pipette

- some paper towels.

Work carefully and sensibly in this activity so you don't make too much mess. Your teacher may suggest that you wear an apron or overall to keep your clothes clean. If anything gets spilt on the floor, tell your teacher at once because it will be slippery.

Put some water in one of the bowls and pour in some oil. Look carefully at what the oil does. Does it sink, mix with the water or float?

Take a close look at the feather. Notice the way it is made and what the surface looks like. Dip your feather into the oily water and swirl it around a bit to make sure that it is covered with oil. Place it on a paper towel and look carefully to see what effect the oil has had on the feather. Can you wipe all the oil off the feather with the paper towel?

Discuss with your partner or group why sea birds might be especially at risk from oil spills.

Use the pipette to drip one or two drops of washing-up liquid into the oil in the bowl and watch what happens. When no more change is happening, add a few more drops of washing-up liquid and stir it carefully with the spoon. You should see how the washing-up liquid helps to break the oil up into smaller droplets.

We use washing-up liquid to help us to clean oily and greasy plates and pans. Do you think this would be a good way to clean up an oil spill in the sea? What problems might it cause?

People who help birds that have become covered with oil often give them a bath in a substance similar to washing-up liquid. This helps to get the oil off the birds' feathers. Try adding a few drops of washing-up liquid to your feather to see if you can wash the oil off.

Pour some water into the second bowl and add some oil as before. Imagine that this is an oil spill in the ocean. Discuss with your partner or group how you might clean up the oil spill. What might you use to help you? (No washing-up liquid or other chemicals allowed this time!) When you have a good idea, tell your teacher about it. Your teacher will give you the things you need to try out your idea.

When you have tried your idea, discuss whether you think it worked well. What was good about your idea? What did not work well? How could you make it better? Compare your idea with those of the rest of your class. Which method of cleaning up the oil was most successful and why? Do you think that this method could be used on a real oil spill at sea?

Habitat destruction

Earlier in this chapter we discussed how the habitat of an animal or plant provides everything it needs to survive.

Humans often cut down forests to make space for growing crops. We build houses in places where there were fields and hedgerows. When we do these things, we are taking away the habitat of all the plants and animals that live there. Sometimes animals are able to move away and find a new home somewhere else but often they cannot.

Humans need space to grow food plants and other crops. We need fields for our farm animals. We need places where we can build homes and other buildings. However, we also need to think about the other living things in the world and try our very hardest to look after their habitats as we build our own.

What actions can we take to make sure that we preserve the habitats of other plants and animals?

◾ These trees have been cut down to make farmland. The rainforest habitat has been destroyed in this place

Endangered species

The more we damage the habitat of a particular type of animal or plant, the more difficult it becomes for it to survive. If animals do not have many places where they can feed, they become weak. If they do not have safe places to breed, they will have fewer babies. Over time there will be fewer and fewer of them.

Some animals and plants are only found in a very small number of places. If we spoil the environment in these special places they cannot survive at all.

When there are very few individuals of a particular type of animal or plant left in the world we say that it has become **endangered**. When this happens, we need to do something to make life better for it so that its numbers can rise again. If we do not, there is the risk that all of the endangered animals or plants will die and the type of animal or plant (the **species**) will not exist any more. When this happens we say that the species has become **extinct**.

◾ Lemurs in Madagascar are endangered due to habitat loss and hunting

Lemurs are endangered animals that live in the forests of Madagascar. The forests are being cut down very fast for farmland and timber. Many people in Madagascar are very poor and they hunt the forest animals, including rare lemurs, for meat. There are many different species of lemur and nearly all of these are now endangered.

Tigers are also becoming very rare due to habitat loss and hunting. In the last 100 years tiger numbers have fallen from hundreds of thousands to around 3500. It is very difficult to count tigers so there could be even fewer than this. There is a very real possibility that tigers might become extinct in your lifetime.

So what can we do to stop this from happening? The first thing is to look at what we have done to its habitat. No animal or plant can survive if it does not have a suitable habitat. There needs to be plenty of places where it can live, preferably all connected together so that they can move from one to another and meet other members of the species to mate. We

Tigers are endangered and may soon become extinct

must make sure that we do not spoil any more of the habitat. We may also need to try to put back some of the habitat that we have taken away, but this is very difficult.

In many places people have set up special areas where wildlife can be protected and helped to thrive. Sometimes these are small, maybe a single pond where rare water creatures live or a small copse that is good for nesting birds. Sometimes they are huge. For example, the government in Sri Lanka has decided to protect all the mangrove swamps along their coast – an area of about 8800 hectares.

All over Britain there are many protected areas, often called nature reserves. These may be looked after by one of the wildlife charities, such as the RSPB, the Woodland Trust or one of the county wildlife trusts. Sometimes they are cared for by small groups of local people.

Did you know?

This picture shows a common otter. This is not a very good name for them because, although there were once lots of otters in Britain, they are not commonly seen now.

There are many reasons why otters have become rare, all due to human activity. Otters eat fish and need clean, fast

moving river water to feed in. They need riverbanks with tall plants and suitable places to build their dens, called holts. In the 1950s and 1960s, many rivers became polluted and the riverbank habitat was spoilt in many places by housebuilding and changes in farming. In some places, otters were killed by fishermen who wanted to protect the fish stocks in their favourite fishing rivers. Otters were also hunted for their thick, soft fur. All these problems together made otters very rare in Britain.

Otters are now beginning to make a come-back. Their breeding sites are now protected and it is illegal to kill them. In places where otters could still be found, people worked hard to improve the riverbanks to help them to breed and spread into new areas. Some otters were bred in captivity and then put back into their habitats. They are still not common but you have a better chance of seeing one in the wild than your parents did when they were your age.

Activity – nature reserves

Find out about a nature reserve near where you live. What kinds of habitat are found in the reserve? What plants and animals can be found there? Are there any special or rare living things that make it their home? Why was this area made into a nature reserve? In what ways does the nature reserve help the plants and animals to thrive? Who looks after it? What could you do to help to keep this place special?

You could make a display about your nature reserve. If you are lucky enough to visit the reserve with your teacher, your family or friends, you could take some photographs for your display. Perhaps you could write a report about your visit saying what you saw and what it was like to be there.

Sometimes, a few of the remaining individuals of an endangered species can be taken to zoos or other places where they can be kept safe. If they are well looked after, they might breed and, when there are enough of them, they can be released into the wild again. This is only possible if there is enough suitable habitat for them to live in.

Exercise 2.2a

1 Describe two ways in which humans have changed the environment near your school.

2 What does the word 'pollution' mean?

3 Describe one way in which human activity might pollute the environment.

4 Name two animals that have become rare because humans have spoilt their habitat.

5 What word do we use to describe an animal or plant when there are very few of them left?

6 How might humans help these animals or plants survive?

7 What might happen to these animals or plants if we do not help them?

Exercise 2.2b

Use the following words to help you to fill in the gaps in these sentences.
Some words may need to be used more than once.

destroyed   endangered   extinct   habitat   pollution   rubbish

1 Human activities may put things into the environment that shouldn't
be there. This is called _____.

2 The _____ of a living thing may be _____ by human
activities such as cutting down trees or dumping _____.

3 When very small numbers of a living thing are left on Earth we say that
it is _____.

4 A living thing is said to be _____ when no more of them exist
on Earth.

5 To help a rare living thing survive we need to look after its _____.

Exercise 2.2c: extension

People often dump rubbish on the roadside – you may see this near
your school. Design a poster to put up in the area explaining why this is
harmful to the environment and to people.

Food chains

In the last chapter you learnt that a habitat is the place where an animal or plant lives. The group of animals and plants living in a habitat is called a **community**. The plants and animals in a community all depend on each other to survive.

All living things need energy in order to live and grow. You learnt in Year 3 that plants need light energy from the Sun, which they use to make their own food using carbon dioxide from the air and water from the soil. Every time an animal does anything, such as run, jump, breathe or keep warm, it uses energy. Animals get their energy from the food they eat.

We can show how living things depend on each other for food in a **food chain**. A food chain shows how energy is passed from one living thing to another.

➲ Producers and consumers

Plants are called **producers** because they produce (make) their own food. They trap energy from the sunlight to make food in the process called **photosynthesis**. Plants are the only living things that are able to do this so all food chains start with a plant.

Animals are called **consumers**. They cannot make their own food so they consume (eat) other plants and animals. This food gives them the energy they need for life.

There are three different types of consumers:

Herbivores are animals that eat only plants. Cows, sheep, giraffes and elephants are all herbivores.

Carnivores are animals that eat other animals. Lions, tigers, sharks and eagles are all carnivores.

Omnivores eat plants and other animals. Foxes, badgers, hedgehogs and most humans are omnivores.

■ Eagles eat other animals. They are carnivores

Did you know?

Scientists who study animals need to find out what the animal eats. For example, they might see a small mammal eating an insect, but how can they tell if this is the only food this animal eats? It is possible to tell a lot about what an animal eats by looking for the leftover bits from its meal. For example, you might find nutshells, bones or the shells of snails. If the animal has eaten a mammal or bird you might find fur or feathers left behind. Often the best way is to study the animal's droppings. Seeds, bones and bits of insect exoskeleton can often be found in droppings.

One type of dropping that can be very interesting to study is owl pellets. Owls eat their prey whole and the bones and fur or feathers get bundled up inside the owl's stomach. It then brings this bundle up into its mouth and spits it out. If you know of a place where owls regularly roost you may find some of these pellets. If you pick one up and pull it apart you can see what the owl ate yesterday. You should wear gloves and remember to wash your hands very well afterwards.

■ This pellet from a barn owl contains the bones of the small mammals it ate

➲ Building a food chain

The first living thing in any food chain is a plant (a producer). This is because plants are the only living things that can trap light energy from sunlight and turn it into food energy.

The next link in the chain will be a herbivore, which eats the leaves, roots, flowers or fruits of the plant to give it energy. The energy from the Sun that was trapped in the plant during photosynthesis will be passed on to the herbivore. The energy will help the herbivore to grow and live a healthy life.

Next in the chain will be another consumer. This will eat the first animal so it will be a carnivore. This carnivore will catch and eat the herbivore. The energy that was used to build up the body of the herbivore is now passed on to the carnivore.

In some food chains a second carnivore will catch and eat the first carnivore. There may be even more carnivores in the chain but this does not happen very often. The last carnivore in the chain is called the **top carnivore** and is usually a large animal.

Remember that an omnivore eats both plants and animals. An omnivore can be found at different levels of a food chain acting as a carnivore or a herbivore.

We draw a food chain by linking the living things together with arrows. When we draw the arrows in a food chain, it is important that they show the direction in which the food energy is travelling. The arrow means 'is eaten by'.

Here is a picture showing the living things making up a simple food chain:

The grass makes its own food. It is a producer so it starts this food chain. The rabbit is a herbivore and gets its energy when it eats the grass. The rabbit is the second link in the chain. The fox eats the rabbit so it is acting as a carnivore. Nothing will eat the fox so it is the last link in the chain.

We show this energy flow like this:

There will be lots of different food chains in a habitat. Here are some food chains that you might find in a woodland habitat.

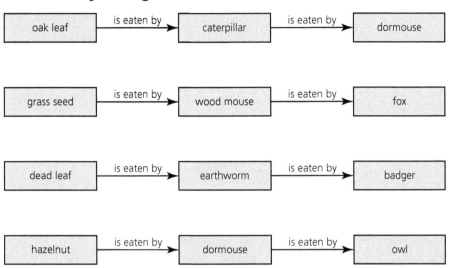

Food chains often change depending on what foods are available to be eaten. For instance, a fox may eat rabbits most of the time but might catch a chicken or a duck if it gets a chance. In the autumn the fox may eat ripe blackberries or other fruits.

Food chains also vary in length.

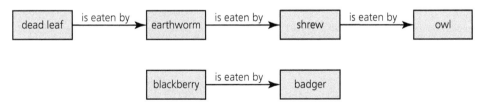

It is important understand the way the different animals and plants in a habitat depend on each other. If we know what eats what, we can work out how changes in a habitat will affect the living things that live there.

In the last chapter you learnt about some ways in which we can try to help living things that have become endangered. To protect an animal we need to know what it eats. We also need to know about all the other food chains in the habitat. For example, dormice are becoming quite rare. We know that dormice eat nuts, and also caterpillars and other small insects. We can look after the nut trees but other plants may also be important to the dormouse because they are food for the small animals that it eats.

Activity – food chains

1 Look at the food chains on the previous page. Work with your partner or group to say whether each of the living things is a producer, a herbivore, a carnivore or an omnivore. Explain your reasons for your choices.

2 Look at this picture. It shows the living things making up a food chain. You can see a caterpillar, a sparrow hawk, some leaves and a blue tit. Draw a food chain, like the ones on the previous page, that includes all these living things.

Exercise 3.1

The table gives some information about the diets of some animals in a pond habitat.

Animal	Foods the animal eats
pond snail	pondweed
stickleback (fish)	dragonfly larvae, water fleas, mayfly larvae
kingfisher	sticklebacks, dragonfly larvae
dragonfly larva	pondweed, mayfly larvae, pond snails
water flea	microscopic plants
mayfly larva	pondweed

1 Use the information to complete the food chains below.

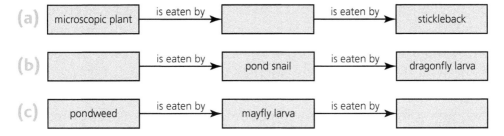

(a) microscopic plant → is eaten by → [ ] → is eaten by → stickleback

(b) [ ] → is eaten by → pond snail → is eaten by → dragonfly larva

(c) pondweed → is eaten by → mayfly larva → is eaten by → [ ]

2 Using only information given in the table, make up a food chain that includes the kingfisher.

⊃ Predators and prey

When an animal catches other animals to eat, we call it a **predator**. The animal that is eaten is called the **prey**.

Predators may have sharp claws, teeth or beaks to grasp their prey. Their eyes are usually on the front of their heads, facing forwards. This allows them to judge distances accurately when they are hunting.

■ A cheetah is a predator. Its front-facing eyes help it to judge distances when hunting

Prey animals often have their eyes on the sides of their heads. This gives them a better view all round so that they can watch out for predators as they feed and move about.

■ A rabbit's eyes are on the side of its head so it can look all around itself for predators

37

Some predators can run very fast but this uses a lot of energy. A cheetah can run at speeds up to 30 metres per second (nearly 110 kilometres per hour or 70 miles per hour). However, it can only run like this for a very short time. Most of its prey animals, such as gazelle, are also fast runners and can dodge around, making it hard for the cheetah to catch them. They can also keep going for longer than the cheetah. This means that most of the cheetah's chases for prey are not successful.

Activity – the food chain game

In this game you will be acting out how predators and prey behave. Some of you will be plants and some of you will be prey animals. A few will be predators. You will pass 'energy' from one to the other as you play the game.

You will need:

- a bowl or other container to represent the Sun

- lots of lolly sticks or strips of card to represent the energy

- bibs or bands in three colours to show which part each person is playing (for example, green for plants, yellow for prey and red for predators).

Your teacher will show you where the limits of the 'habitat' are. No one must run outside the habitat.

Place the 'Sun' bowl in the middle of the habitat and put the 'energy' strips or lolly sticks into it.

Choose a few people to be plants. These people should put on the plant colour bibs or bands, spread out around the habitat and stand still. Each plant should start with a few energy strips. When a plant runs out of energy, they can go to the Sun and collect a few more pieces of energy. Your teacher will tell you how many you can take.

Choose two people to be predators. These people should put on the correct colour bibs or bands and stay at the side of the habitat to start off with.

The rest of you are prey. Put on your bibs or bands and wait for the teacher to say 'Go'.

When the teacher says you can start, the prey animals should start to collect energy from the plants. If a prey animal comes up to a plant the plant must give them one piece of energy. A prey animal cannot take more than one piece of energy from a plant at one time but you can go back to that plant later when you have visited some others. The prey animals should try to collect as much plant energy as possible.

After a short time, the teacher will tell the predators to start. The predators want to get energy from the prey animals so they should chase them. When a predator touches a prey animal, the prey animal is dead and is out of the game. The energy that the prey had collected goes to the predator.

The game finishes when there are no more prey animals left. The predators can count their pieces of energy to see which of them was most successful.

You can now play the game again but change the number of predators. What happens if there are four predators instead of two? How much energy does each one get? What happens to the length of the game before all the prey animals are eaten? Use your understanding of food chains to help you to explain the changes that you observe.

What other questions might we ask about changes in the food chain? Discuss these with your class and then change the game to help you to answer your questions.

Exercise 3.2a

1 Explain why nearly all food chains start with a plant.

2 What do plants need to make their food?

3 What is the difference between a herbivore and an omnivore?

4 What do the arrows in a drawing of a food chain mean?

5 What is a predator?

6 Why do predators generally have their eyes on the front of their heads?

7 Antelope have their eyes on the sides of their heads. What does this tell you about their position in a food chain?

Exercise 3.2b

Use the following words to help you to fill in the gaps in these sentences. Some words may need to be used more than once.

animals   carnivores   consumers   energy   food   herbivore

photosynthesis   plants   producers

1 In a food chain, plants are described as _____ because they can make their own _____ using the process called _____.

2 Animals need food to provide them with _____.

3 In a food chain, animals are described as _____.

4 A _____ is an animal that eats plant material.

5 Animals that only eat other animals are called _____.

6 Omnivores eat _____ and _____.

Go further

Some animals feed on the bodies of animals that have already died. We call these animals **scavengers**. Examples of scavengers are magpies, crows and red kites, which may be seen eating the remains of animals that have died on the roads. Vultures are also scavengers. Large carnivores, such as lions, will also scavenge dead animals if they can because it is easier than catching the prey for themselves.

🔲 Vultures are scavengers

Scavengers play a useful role in clearing away the bodies of dead animals quickly. Without them we would have to wait for the bodies to break down slowly. This would be done by tiny living things called **decomposers**. Decomposers feed on the dead material and break it down into smaller parts. This helps to return minerals to the soil where they can be taken up by plants. The decomposers that feed on dead bodies are mostly bacteria and fungi. You will learn more about these in Year 6.

Activity – make a food chain mobile

You will need:

- some pieces of card
- a hole puncher
- treasury tags.

You will also need a food chain to make into a mobile. You could choose one of the ones from this chapter. If you have been learning about a different habitat, for example a rainforest or a desert, you could choose a food chain from this habitat.

To make your food chain mobile, you will put each link in the chain onto a separate card. You should write the name of the living thing onto the card and then draw a picture or maybe print out a picture, cut it out and stick it onto the card.

On the back of each of the cards, write whether the living thing on that card is a producer, a herbivore or a carnivore.

Make another card for the Sun because this is where all the energy comes from in the first place.

Use the hole puncher to make a small hole at the top and bottom of each card.

Starting with the Sun at the top, use the treasury tags to join all the cards together in the right order for your food chain.

You can then hang your food chain up in your classroom or at home to remind you about food chains.

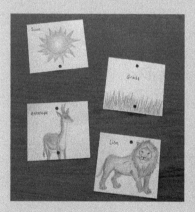 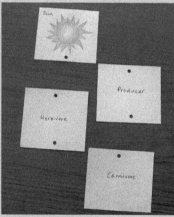 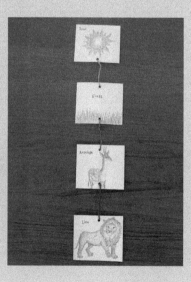

Go further

➲ Plants can consume animals too!

Some plants live in areas where either the soil is very thin, or it is of poor quality and does not contain many minerals. Some of these plants have developed a special trick. They still carry out photosynthesis to make food but they also catch insects and other small animals, and digest them. In this way they take in the minerals they cannot get from the soil.

One kind of carnivorous plant is the pitcher plant. Some of its leaves have changed so they are rolled into a deep cup-shape. The sides of the cup are very slippery and there is liquid in the bottom. Insects that land on the trap slip down into the cup and drown in the liquid. The liquid slowly digests the insects and the plant absorbs the minerals that are released.

■ Insects fall into the trap of a pitcher plant and provide the plant with minerals

Another carnivorous plant is the Venus flytrap, which has sensitive hairs on special leaves. When an insect walks over the leaf, it touches the hairs and triggers the leaves to snap shut, trapping the insect inside. The leaf then releases chemicals that digest the insect.

There are other carnivorous plants which have sticky leaves that trap insects, and some others which have

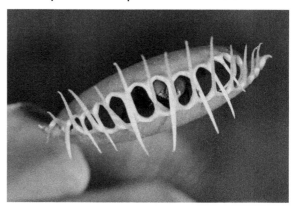

■ The leaves of the Venus flytrap snap shut to trap insects

a one-way route, with forward pointing hairs that stop the insect from turning around and escaping.

Charles Darwin was a famous British scientist. You will learn more about him in Year 6. He asked many questions about the living things that he observed in his travels around the world. He carried out many experiments using plants in his garden and greenhouse to answer some of his questions.

He wanted to find out whether carnivorous plants could eat everything that landed on their traps. He tried feeding Venus flytraps all sorts of different foods. He gave them tiny pieces of foods such as meat, egg or fruit, and drops of sugar solution. He found that the flytraps closed quickly around foods that were high in protein but did not close at all round the sugar solution. It seems that Venus flytraps can detect what is good for them!

■ Charles Darwin did experiments to answer his questions about Venus flytraps

Exercise 3.3: extension

1 Explain why some plants need to catch insects to remain healthy.

2 Name two different types of carnivorous plant.

3 Describe how one of the plants you have named in Question 2 catches insects.

4 (a) Suggest what question Charles Darwin was trying to answer when he did his experiments with Venus flytraps.

   (b) What did he learn from his experiments?

Teeth and digestion

In the last chapter you learnt about the different diets eaten by animals. Animals need to gather their food in some way. Herbivores need to bite through twigs and leaves. Carnivores need to catch their food. Many animals chew their food into smaller pieces to make it easier to swallow. All these activities are made easier by the right kinds of teeth.

⊃ Different kinds of teeth

There are four main types of teeth: **incisors**, **canines**, **pre-molars** and **molars**. Each of these is specially shaped to do a particular job.

Incisors are broad, thin teeth, shaped like a knife blade. They are useful for biting and cutting through foods such as grass. They are also useful for gnawing away at tougher materials, such as wood. Incisors are found right at the front of the mouth of the animal.

■ Beavers have large incisors to help them cut through wood

Canines are sharp pointed teeth. They are useful for tearing meat. They are also very good at catching hold of moving prey. Predators have large, sharp canines but many herbivores do not have any canine teeth. Canines are found just behind the incisors in animals that have them.

■ The lion's long, sharp canines help it to catch prey and to tear the meat

Pre-molars and molars are broad teeth. They have a surface covered with ridges or bumps. They are used for grinding or chewing food to make it easier to swallow. Some animals, such as elephants, have huge molars to help them to grind up the tough plant material that forms their diet.

 Exercise 4.1

The table shows the different types of tooth and their **functions** (jobs). There are some gaps in the table.

Copy the table and then use the information above to help you to fill in the gaps.

Type of tooth	Function
	biting and cutting
molars and pre-molars	
	tearing and catching prey

⬈ Be an animal teeth detective

We can tell a lot about an animal's diet by looking at its teeth. Your teacher may be able to show you some real animal skulls to look at. There are also some photographs below.

A

B

C

Take a close look at the skulls (real or photographs). Look especially at the teeth. Discuss with your partner or group what types of teeth you can see.

Make a list of the teeth in each of the skulls. What do you think the arrangement of teeth tells you about the diet of the animals and how they get their food? What questions would you like to ask about what you can see? Write your questions down and then discuss them with your partner or group. Can you suggest answers to the questions from your own knowledge? If not, use books or the internet to help you to answer them.

The three skulls in the photographs above are of a shark, a horse and a tiger. Which is which? Discuss you ideas with your partner or group.

 Did you know?

Some animals, such as elephants, have teeth that keep on growing all through their lives. The rough plant material that elephants eat wears away the teeth. If the teeth did not keep growing, the elephant would no longer be able to chew up its food so it would starve.

■ Elephants' teeth keep growing all through their lives

Sharks regularly lose their teeth and grow new ones. Some sharks will replace their teeth about every two weeks. How do you think this helps the shark to feed?

➔ Human teeth

Humans eat a wide variety of foods. We are omnivores. This means that we need all four types of teeth. Incisors at the front of the mouth cut our food into small enough pieces to fit into our mouths. Sharper canines help to tear at tougher foods. At the back of the mouth, we have pre-molars and molars to chew the food and make it soft enough to swallow.

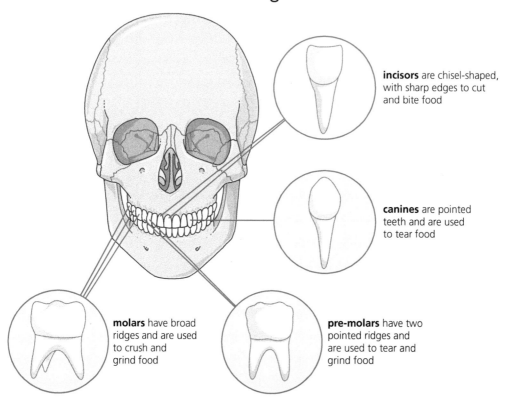

incisors are chisel-shaped, with sharp edges to cut and bite food

canines are pointed teeth and are used to tear food

molars have broad ridges and are used to crush and grind food

pre-molars have two pointed ridges and are used to tear and grind food

Human babies are born without teeth. When they are about five or six months old their first teeth begin to show through their gums. These are part of their first set of teeth, often called 'milk teeth'. It usually takes about three years for all the milk teeth to come through.

Babies' jaws are not very big so there are only 20 milk teeth. When the child is a bit older, usually about five or six years old, these first teeth begin to fall out and are replaced by new, adult teeth. When all these adult teeth come through, there will be 32 of them.

Humans have only two sets of teeth, the milk teeth and the adult teeth. Your adult teeth have to last you for the whole of the rest of your life so it is important that you look after them carefully.

Human teeth

You will need:

- colouring pencils

- a mirror.

Here is a tooth chart showing all the teeth that you will have when your adult teeth have all come through. You will need a copy of this. Your teacher may give you one or you can copy it yourself.

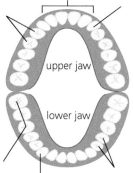

our teeth

upper jaw

lower jaw

Working with your partner or group, use the picture of the human skull and teeth on the previous page to help you to identify and label the different types of tooth in this tooth chart. Check your ideas and labels with your teacher. When you have got them all right, colour each type of tooth a different colour.

Copy this table and complete it, using your tooth chart.

Type of tooth	Number in top jaw of adult	Number in bottom jaw of adult	Total
incisors	4		
canines		2	
pre-molars		4	
molars	6		

Now use the mirror to look at your own teeth. Identify the different types of teeth in your mouth. How many teeth do you have altogether?

Mark on the tooth chart which adult teeth you already have. You could also mark any that are wobbly. Which milk teeth did you lose first? Is this the same for everyone?

➲ Looking after your teeth

A tooth is made up of two parts: the **crown** and the **root**. The part that you can see, outside the gum, is the crown. The other part, which sits inside the gum and holds the tooth in place, is called the root. Unless you have a tooth removed by the **dentist**, you will never see the roots of your teeth. When your milk teeth fall out, the root has been dissolved away so you only see the crown.

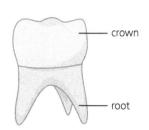

crown

root

■ A tooth has a crown and a root

If you could look inside one of your teeth, you would see that it is made up of three layers. The outside layer of the crown is called the **enamel**. Inside the enamel is a softer layer called **dentine** and in the middle is an area that contains blood vessels and nerves. This is called the **pulp**.

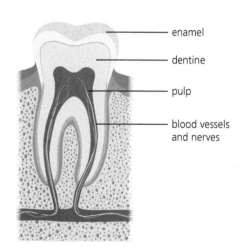

enamel

dentine

pulp

blood vessels and nerves

■ The inside of a molar

To keep your teeth healthy, you need to make sure that the hard enamel layer is not damaged. Sometimes teeth are damaged by accident but most damage is caused by tiny living things, called bacteria, which live in your mouth. If you do not clean your teeth properly, these bacteria form a layer on the surface of the tooth. We call this layer **plaque**.

Tooth decay

Bacteria feed on sugar. When you put sugary foods or drinks in your mouth, the bacteria begin to feed on the sugars. As they do so, they turn the sugar into acid. If the acid is not cleaned off, it can eat away at the enamel surface of the tooth. This might make a small hole in the enamel, called a **cavity**.

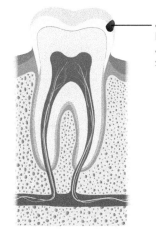

a cavity caused by acid eating away the enamel surface

■ A hole in the enamel of a tooth is called a cavity

This cavity is a good place for bacteria to live because you cannot brush them away easily. They continue to feed on sugars, making more acid. The cavity gets bigger and bigger.

If the dentist does not find the hole and mend it, the acid will wear away the soft dentine. Soon it will get right through to the pulp where the nerves are, causing pain. Ouch!

To stop this happening you need to try to keep the bacteria away and make sure that they are not too well fed.

The most important thing is to make sure that your teeth are properly cleaned at least twice a day. You can use an electric toothbrush or a hand one, but make sure that it is in good condition. You should use a small amount of toothpaste, about the size of a pea. Many toothpastes contain a mineral called **fluoride**. This makes the enamel coating on the crowns of your teeth stronger.

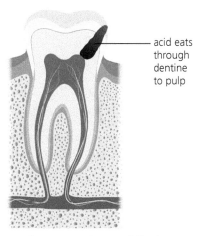

acid eats through dentine to pulp

■ If a cavity is not filled, it will get bigger

When you clean your teeth you should scrub them gently for about two minutes. This seems like a long time but it is the best way to remove all the plaque. Make sure that you scrub each tooth, especially the places where bacteria can get stuck, such as between the teeth and where the tooth meets the gum.

Many people clean between their teeth using dental floss or special little brushes. Your dentist will show you how to do this properly.

Activity – clean and shiny teeth

How good do you think you are at cleaning your teeth? This activity will show you.

You will need:
- your toothbrush and toothpaste
- a mirror
- a disclosing tablet.

Disclosing tablets stain plaque. They can show you where the plaque is on your teeth so that you can learn how to clean them really well. They can also stain your clothes, so be very careful when doing this activity.

Do this activity near a clean basin with fresh running water.

Start by chewing the disclosing tablet and swilling it around your mouth for about a minute. Carefully spit the remains out into the basin.

Now use the mirror to look at your teeth. Any areas of plaque will be stained, usually a pink colour, by the disclosing tablet.

Put the mirror down and use your toothbrush and toothpaste to clean your teeth as you usually would. Make sure that you leave the basin clean when you have finished.

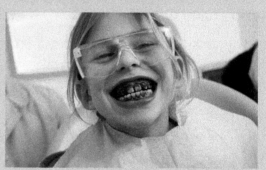

A disclosing tablet shows where the plaque has built up on your teeth

When you have cleaned your teeth, have another look at them using the mirror. Have you cleaned off all the plaque or are there still some places where you can see the stains? Take note of the places where you did not manage to clean all the plaque off. You need to make sure that you take care to brush these places whenever you clean your teeth. Your teacher may let you have another go at cleaning the rest of the plaque off so that your teeth are all shiny and clean.

When you have finished this activity, make sure that your toothbrush is put somewhere where it will be kept clean until you need it next.

Fight the bacteria

Cleaning your teeth properly twice a day is very important. You can also help to keep your teeth healthy by not giving the bacteria very much to feed on. Try not to eat too many sugary foods. Sweet, fizzy drinks are really bad for your teeth because they contain acid as well as sugar.

You should also visit your dentist regularly. Most dentists suggest that you should have your teeth checked about twice a year. The dentist will check your teeth for cavities. If your dentist finds any cavities, they can mend them before they get any bigger. Your dentist may also give your teeth an extra scrub in some places to get them really clean.

Exercise 4.2a

1 Name the four main types of teeth.

2 What is the name of the part of the tooth that can be seen outside the gum?

3 What is enamel?

4 How many milk teeth do children have?

5 How many teeth do adults have?

6 What is the name given to the layer of bacteria that builds up on our teeth?

7 Explain in your own words how bacteria cause the teeth to decay.

8 What three things can you do to make sure that your teeth stay healthy?

Exercise 4.2b

Use the following words to help you to fill in the gaps in these sentences. Some words may need to be used more than once.

acid   adult   bacteria   canines   cavities   enamel   incisors   milk

molars   pre-molars

1 There are four different kinds of teeth. They are called _____, _____, _____ and _____.

2 When children are about five, their _____ teeth begin to fall out.

3 These are then replaced by _____ teeth.

4 If we do not clean our teeth, _____ will turn sugar into _____ , which will damage the _____ and make _____.

Exercise 4.2c: extension

1 Suggest a reason why human babies are born without teeth.

2 Imagine that you are a bacterium on the teeth of a young child. Write a diary entry for a day. Include details about what the child eats and what happens when the child cleans his or her teeth and visits the dentist.

Activity – persuasive writing

Most of us know that sweet foods are bad for our teeth. We also know that these foods are tasty and we like to eat them. Dentists want to persuade people to eat fewer sugary foods and to clean their teeth properly.

Make a leaflet that the dentist might give out to children of about your age. It should explain clearly why they should not eat too many sugary foods. It should also tell them how to clean their teeth properly and why it is necessary.

Your leaflet should be clear, attractive and persuasive. You can include pictures to help you achieve these objectives.

Did you know?

Have you ever wondered what people did before toothbrushes and toothpaste were invented? Ancient people would not have known about bacteria and how they cause tooth decay. However, it seems that many of them did clean their teeth in some way.

In the Stone Age some people probably used small twigs or pieces of bone to remove bits of food that had got stuck between their teeth. It seems that some chewed on twigs or roots of herbs such as mint or ginger, which would have freshened the breath as well as cleaning the teeth.

It is known that the Ancient Egyptians and the Chinese had some kind of paste that was rubbed on the teeth to clean them. The recipes for this included some strange things, such as ground ox hoof, honey, salt and ground bones! This was probably rubbed onto the teeth using a brush made from plant stems with frayed ends. The Ancient Greeks even had a sort of chewing gum, called mastiche, which they chewed to clean their teeth.

The Chinese had toothbrushes with bristles by the fifteenth century and toothpaste as we know it was probably invented in the nineteenth century.

◔ Food in the body

Teeth are very useful for gathering and breaking up food. The food is then swallowed and 'disappears' down into the body. What happens to it next?

Inside your body is a group of **organs**, called the **digestive system**. You learnt a bit about them in Year 3. Here is a diagram of the inside of the human body, showing these organs.

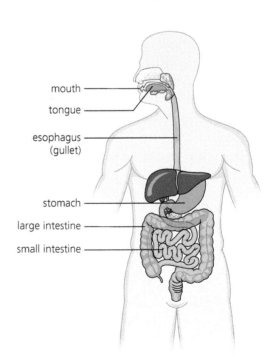

mouth

tongue

esophagus (gullet)

stomach

large intestine

small intestine

Look at the diagram of the digestive system on the previous page. Your teacher may also show you a model of the body with all the organs visible.

In your notebook, write down what you think happens to the food when it is swallowed. You can draw some pictures if you like, to help. Discuss your ideas with your partner or group. Do you all think the same or do your ideas differ? If you disagree, explain your ideas clearly to each other and then discuss what seems good about each one. Say politely if you think someone is wrong and explain why.

When you have finished your discussion, make any changes that you think would improve your work.

When you put some food into your mouth you prepare it to be swallowed. Your tongue moves it around in your mouth to make sure it is all chewed properly. It is also mixed with a special liquid called **saliva**. This makes the food softer and easier to swallow.

When you swallow your mouthful, it goes down the long pipe, called the **esophagus** or **gullet** and into your **stomach**. It then stays in your stomach for a few hours. Inside the stomach it gets mixed with some special chemicals, which break most of the food down into very tiny pieces. This is called **digestion**.

When the food has been digested it moves from the stomach into another, very long tube, called the **small intestine**. In this place some of the very tiny pieces of food can move from the digestive system into the blood to be carried round to all parts of the body. This is how the **nutrients** in the food reach the cells in your body. You will learn more about this in Year 6.

Some parts of the food are not well digested in the human body. They cannot be taken into the blood in the small intestine. Most of this is the fibre that comes from plant material in your diet. You learnt about fibre in Year 3.

Fibre does not give you nutrients for energy or for growth and repair. It does not give you the vitamins and minerals that you need to stay healthy. However, fibre is important, because as it moves through your guts, it helps to keep your digestive system healthy. It passes right through the small intestine and into the **large intestine**. In this part of your body most of the water is taken out of what is left from your food and it is parcelled up into lumps and stored until you are ready to let it come out of your body when you go to the toilet.

Activity – share your ideas: part 2

Now you have read about the digestive system and discussed it with your teacher, go back to the ideas that you recorded in your book earlier. Were your first ideas correct? If not, what did you get wrong? Imagine that you are a teacher. Using a different coloured pen or pencil, mark your work to show the changes that need to be made to correct the work.

Exercise 4.3a

1 The digestive system is made up from these organs: small intestine, mouth, esophagus (gullet), large intestine, stomach.

   Write the names of these organs in the order in which the food would pass through them in the body.

2 The functions (jobs) of the different organs are given below. In each case, name the organ that does the job.

   (a) digests food

   (b) carries food from the mouth to the stomach

   (c) chews up the food

   (d) takes the digested food into the blood

   (e) takes water out of the waste material that is left

Exercise 4.3b: extension

Draw a strip cartoon, like you might find in a comic book, showing the different stages of the work of the digestive system. You could invent some cartoon characters to represent the different nutrients – those that are taken into the blood and those that get left behind.

5 States of matter

→ States of matter

Everything around us is made from materials that are in one of three states: **solid**, **liquid** or **gas**. We call these states the three **states of matter**. You are probably quite good at telling whether something is a solid, a liquid or a gas but have you ever thought about *how* you can tell?

Activity – solid, liquid or gas?

Look at the pictures below. Discuss with your partner or group whether the material is a solid, a liquid or a gas. Try to say what it is about each one that helps you to decide.

| Air in a balloon | Log of wood | Milk | Ketchup |

Did you find it easy to decide which state each of these materials is in? Was there one that was harder than the others? If so, why was it harder?

➲ Properties of matter

When you were deciding whether materials were solids, liquids or gases, you were thinking about the ways in which materials behave. These are their properties.

When you are deciding the state of a material, three properties are important:

- whether the material can flow
- whether the material can be squeezed
- whether the material keeps its shape and its volume (size) if it is moved about.

➲ Investigating the properties of matter

For these activities you will need:

- three balloons: one filled with air (a gas), one filled with water (a liquid), one filled with water and then frozen so it contains ice (a solid)
- sticky tape
- a large bowl
- a pin
- a pair of scissors.

Activity 1: description snowball

Look at the three balloons carefully. Feel them with your hands (gently!).

Work with your partner to write down as many adjectives as you can to describe each one. Remember that you are describing the materials in the balloons, not the balloons themselves.

Working Scientifically

Now join up with another pair and compare your words. Add to your list by writing the words that the other group thought of but you did not.

If you have time, you can then join with another group of four, and then maybe with another group of eight. How many good adjectives can the class think of to describe the materials?

Activity 2: escaping materials

In this activity you are going to investigate whether each of the materials can flow, keep their shape and keep their volume (size).

Take two pieces of sticky tape, each about 5 cm long. Stick them onto the balloon containing air to make a cross shape. Make sure they are firmly stuck on.

Now take the pin and very gently push it through the place where the two strips of tape cross. You should be able to make a small hole right through without popping the balloon.

Pull the pin out very gently. Take it in turns to hold the back of your hand near the hole. Can you feel the air escaping? You might find it easier to feel if you make the back of your hand damp first.

Discuss what you can feel with your partner or group. Is the air able to flow? Where is the air going to? Has the air kept its shape? If not, what shape is it now? Has the air kept the same volume (size)?

Do you think all gases would behave in this way?

Now take the water balloon. Put it in the bowl and carefully use the scissors to cut off the knot. Watch what happens to the water when you let go of the balloon.

Discuss what happens with your partner or group. Is the water able to flow? Where has the water gone to? Has the water kept its shape? If not, what shape is it now? Has the water kept the same volume (size)?

Do you think all liquids would behave in this way?

Remember to wipe up any spilt water, especially if it is on the floor.

Next empty the water out of the bowl and put the ice balloon in the bowl.

Carefully take the balloon off the ball of ice. You may find that it has already split but, if not, use the scissors carefully to remove it.

Discuss what you can see with your partner or group. Is the ice able to flow? Has the ice kept its shape? If not, what shape is it now? Has the ice kept the same volume (size)?

Do you think all solids would behave in this way?

Activity 3: the big squeeze

You are going to find out whether solids, liquids and gases can be squeezed to make them smaller. If a material can be squeezed in this way, we say that it is compressible.

For this activity you will need:

- your ball of ice or an ice cube
- a syringe
- some water in a beaker
- a bowl.

Take the syringe and pull the plunger about halfway out so that the syringe is about half full of air.

Press your finger tightly onto the nozzle so no air can escape and then try to push the plunger back in.

Can you squeeze the air and make it smaller? Is the air compressible?

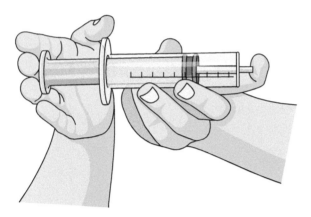

Now take your finger off the end of the syringe and push the plunger right back in. Put the nozzle into the beaker of water and then pull the plunger slowly out until the syringe is about half full of water.

Place your finger over the end again and hold the syringe over the bowl with the nozzle pointing downwards. Try to push the plunger in like you did with the air.

Can you squeeze the water and make it smaller? Is the water compressible?

Discuss with your partner or group what differences you notice between the air and the water.

Lastly, push down on your ball of ice or ice cube with your hand.

Can you squeeze it and make it smaller? Is the ice compressible?

Copy the table on the next page into your book. Use the results of your experiments to help you to fill in the table by writing 'Yes' or 'No' into each of the squares.

Property	Gas	Liquid	Solid
Can it flow?			
Does it keep its shape?			
Does it keep its volume (size)?			
Is it compressible?			

Exercise 5.1

Use the following words to help you to fill in the gaps in these sentences. Some words may need to be used more than once.

compressible   gas   gases   liquid   liquids   shape   solid   solids volume (size)

1 The three states of matter are _____ , _____ and _____.

2 If a material can be squeezed to make it smaller we describe it as _____.

3 A _____ will keep its shape when moved about.

4 Both _____ and _____ can flow.

5 _____ spread out to fill the whole container or room.

6 A _____ takes the shape of the bottom of whatever container it is in.

7 Solids and liquids keep the same _____ when moved from one container to another.

It is easy to tell whether air, water and ice are solids, liquids or gases. Sometimes it is much harder to tell which state a material is in. In the following activity you can use what you have learnt to decide whether the materials are solids or liquids.

→ Mystery materials

Mystery 1: sand

For this activity you will need:

- a beaker of dry sand
- an empty beaker
- a syringe
- some water
- a stirring rod or spoon.

Pour the dry sand from one beaker to the other. Does it flow? Does it keep its shape? Does it keep its volume?

Pull the plunger out of the syringe. Put your finger over the nozzle and pour sand in until the syringe is half full.

Push the plunger a little way in and then hold the syringe with the nozzle facing upwards and take your finger off the nozzle.

Carefully push the plunger in, pushing the sand up the syringe until all the air has been pushed out.

Put your finger back on the nozzle and try to compress the sand. Is it compressible?

Look back at the table you made at the end of the last activity. What properties does the sand have that make it like a solid? What properties does it have that are like a liquid? Do you think the sand is a solid or a liquid? Discuss your ideas with your partner or group. Remember to explain your ideas clearly. Do you all agree?

Add a little water to the sand and mix it carefully. How does this change the sand? Does this make a difference to whether you think it is a solid or a liquid?

Mystery 2: cornflour

For this activity you will need:

- a beaker containing some cornflour
- a beaker containing water
- a pipette
- a stirring rod or spoon
- a shallow dish or saucer.

Add water to the cornflour, a little at a time, using the pipette. You need to add enough to make it wet enough to stick together but not so much that it becomes really sloppy.

Look at the way the wet cornflour behaves when you stir it slowly and when you try to stir it quickly.

Tip the beaker to turn the cornflour into the shallow dish then use your hands to form it into a ball. Place the ball in the dish and watch what happens.

Discuss your observations with your partner or group. Think about the properties of the damp cornflour. Would you say it is a liquid or a solid?

⊙ Particles

All materials are made from very tiny **particles** (pieces) called atoms and molecules. These are so small that we cannot even see them with a microscope. However, the way they behave is important because scientists believe this is what gives solids, liquids and gases their properties.

Particles in a solid

If you could see the particles in a solid, you would see them all packed neatly together, like lots of oranges in a box. They would not be completely still. The particles in a solid **vibrate** (jiggle) as if they were shivering.

Particles in a solid are quite tightly stuck together and cannot move around. This is why solids cannot change their shape and do not flow. As the particles are so close together they cannot be squeezed any closer, so a solid is not compressible.

■ Particles in a solid are packed neatly together

Particles in a liquid

The particles in a liquid are still quite close together but they are not so neatly packed. They vibrate more than particles in a solid and, if you could see them, you might see them move around a bit.

The particles in a liquid are not as strongly stuck together as the ones in a solid so they can change their positions. This means that the liquid can flow and change its shape. The particles are still close together so they cannot be squeezed any closer. A liquid is not compressible.

■ Particles in a liquid are not packed together neatly and can move around a bit

Particles in a gas

In a gas, the particles are not stuck together at all. They are spaced widely apart. They have lots of energy so they are moving around quite fast in all directions and bumping into things. This means that they spread out and go anywhere they can.

The spaces between the particles make it possible to squeeze a gas and bring the particles closer together. This means that a gas is compressible.

■ Particles in a gas can move around freely in all directions

Did you know?

Even when compressed, the particles in a gas still have a lot of energy and so the gas pushes hard on the container it is in. For example, if the gas is compressed in a balloon, the energy makes a loud pop if the balloon bursts and the gas rushes out of the hole.

We can use this energy to move things. The push of compressed gas can be so strong that it can be used to press hard enough on the brakes on a heavy lorry to make it stop.

A bursting balloon releases a lot of energy

Exercise 5.2a

1 Describe in your own words how scientists believe the particles are arranged in a solid.

2 Draw two diagrams to show the arrangement of particles in a liquid and a gas.

3 Explain why gases can be compressed (squeezed) but liquids and solids cannot.

Exercise 5.2b

Choose the correct word or phrase from each pair in brackets to complete the following sentences.

1 In a (liquid/solid) the particles are arranged in a regular pattern.

2 In a gas the particles are (close together/far apart).

3 The particles in a solid (stay still/vibrate).

4 The particles in a (liquid/solid) can move around and change places.

Exercise 5.2c: extension

The first bicycles had wooden wheels with a strip of metal around them to stop them wearing out. Later bicycles had metal wheels and rubber tyres filled with air. Give as many reasons as you can why these later bicycle-wheel designs are better than the old wooden ones.

→ Changes of state

Many materials can exist in more than one state. For example, water is usually a liquid but if you make it very cold, it becomes a solid, called ice.

■ Antarctic iceberg. The liquid water has turned into solid ice

If you heat water up it turns into a gas, called water **vapour**.

As water vapour cools it turns to tiny drops of liquid water that we can see and is called steam.

Water vapour is an invisible gas.

■ When water is heated it turns into a gas called water vapour

It is easy to change the state of water by heating or cooling it. It is also possible to change it from one state to another and back again. If we warm an ice cube it will change into liquid water. If we cool water vapour it turns into drops of liquid water. We describe these changes as **reversible**.

The changes of state have special names.

Freezing and melting

If we take water or another liquid and cool it down enough, the particles become more neatly arranged and stick more closely together. The liquid turns into a solid. This change is called **freezing**.

To make a liquid material freeze completely, we have to cool it down to a special temperature called its **freezing point**. Different substances have different freezing points. The freezing point for water is 0 °C.

When we warm solids up, some of them will turn from solid to liquid. The particles become more energetic and they become less tightly stuck together. This change is called **melting**.

Ice in a freezer is usually at a temperature of about −20 °C. If we warm it up to above 0 °C it will melt to form liquid water. We could say that 0 °C is also the **melting point** of water. Different solids melt at different temperatures.

When water turns into a solid, we call the solid 'ice' but most materials do not have special names for their different states.

Did you know?

Chocolate melts at 37 °C. This is also the temperature of the human body and is why chocolate melts in your mouth so easily.

We can make some delicious treats by changing the state of some foods. Here are two recipes. Remember to wash your hands before working with food and use clean bowls and other kitchen equipment.

Recipe 1: fruity snow

You will need:

- a large bowl
- lots of ice
- salt
- a plastic food bag
- plastic clip or twist/tie
- fruit juice or smoothie
- a spoon.

Pour some fruit juice or smoothie into the plastic bag and seal it up very carefully using the clip or twist/tie so that the juice cannot leak out.

Put lots of ice into the big bowl and sprinkle some salt over the ice.

Bury your bag of fruit juice in the salty ice and wait for it to freeze.

When it has frozen (possibly around 45 minutes – 1 hour) wipe the outside of the bag with a clean cloth and open it carefully.

Look carefully at the fruity snow in the bag. What changes have taken place while it was in the ice bowl?

Now you can use the spoon to sample your fruity snow.

Recipe 2: chocolate fruity cups

You will need:

- chocolate
- raisins or other dried fruit
- a heatproof bowl
- a large bowl or saucepan containing very hot water (be very careful!)
- a spoon
- paper cake-cases.

Break up the chocolate and put it in the heatproof bowl. (Don't eat it yet!)

Very carefully stand the bowl in the hot water.

Stir the chocolate carefully and the heat from the water will gradually melt all the chocolate. (Take care not to get any water into the melting chocolate.) How does the chocolate change? What words could you use to describe the differences between solid chocolate and liquid chocolate?

When the chocolate has melted, stir in the fruit.

Put spoonfuls of the mixture in the paper cases and wait for the chocolate to cool and become solid. When the chocolate is hard, the fruity cups are ready to eat.

Did you know?

When water freezes it **expands** (which means it gets bigger). If you leave water in a hosepipe in freezing conditions, the water can burst the pipe as it turns into ice and expands. You might not notice this until you try to water your garden the next summer and the water leaks out through the split in the pipe.

Evaporation and condensation

Have you ever wondered what happens to the water in wet clothes when they are hung up to dry, or where the water comes from when a window 'steams up'? These things happen because of water changing state from liquid to gas and back again.

When clothes have been washed, some of the water stays in the fabric and needs to be removed before we can wear the clothes again. There are two ways we can do this. We may use an electric tumble dryer, which blows warm air through the clothes, or we may hang the clothes

■ Wet clothes will dry if hung on a washing line

up and wait for the water to disappear. But where does the water go?

The answer is that the liquid water in the clothes turns into the gas called water vapour. The water particles escape from the clothes and mix with the other gases in the air. This change, from a liquid to a gas, is called **evaporation**.

Water can evaporate slowly at any temperature but it happens more quickly if you warm it up. This is why tumble dryers use warm air.

If you make water hot enough, it will begin to **boil**. This means that bubbles of vapour (gas) start forming in the liquid water. They rise to the surface of the liquid and escape into the air. You will also see tiny droplets of very hot liquid in the air. We call this **steam**.

Each liquid will boil at a particular temperature. We call this temperature the **boiling point** of the liquid. The boiling point of water is 100 °C.

■ The boiling point of water is 100 °C

➲ Wash day

Mrs Mop is going to wash her clothes. She wants to know the best place to hang them to dry because she does not have a tumble dryer.

She would like you to carry out an investigation to find out where she should hang her clothes. You do not need to use real clothes for this. You can use squares of fabric or kitchen cloths.

You will need to make your investigation a **fair test**. This means that you will change just one thing – the place where the clothes are hung – and keep everything else the same.

Start by discussing what places you might choose to hang the wet clothes. Look around the classroom and outside and choose three or four different places to try. Think about how the conditions are different in each of the places. For example, are they warm or cool, windy or still? Discuss with your partner or group how quickly you think clothes would dry in each of the places. This is your **prediction**.

In each place, use a thermometer or datalogger to measure the temperature of the air. Write this down in your notebook. You should also record any other conditions that might make a difference, for example is it windy or still, sunny or shaded?

You will need one piece of cloth for each of the places you have chosen. To make the test fair, these need to be the same fabric and the same size.

Decide how you will hang the cloths up in the different places. What will you need to do to make it fair?

You now need to make your pieces of cloth wet. Discuss with your partner or group how you will try to make sure that they are all equally wet.

Hang your wet cloths in the places you have chosen and make a note of the time. You will need to check them frequently to see when each cloth is dry. When a cloth is dry, make a note of the time and work out how long it has taken to dry.

Record your results in a table like the one below.

Place	Air temperature, in °C	Other conditions	Time taken for cloth to dry, in minutes

Draw a bar chart to show the time taken for the cloth to dry in each place.

How will you know which place is the best for Mrs Mop to dry her clothes?

Write a letter to Mrs Mop, explaining how you have done your test and advising her about where is the best place to hang her clothes and why.

Water vapour (gas) can get into the air in lots of ways. Every time you breathe out, you add water vapour to the air because your breath is moist. Plants give out water from their leaves too.

After rain, puddles evaporate and water is always evaporating from seas, lakes and rivers. We cannot see it, but there is always water vapour in the air.

If the air is cooled down, the particles in it slow down. Some of the water vapour in the air will turn into droplets of liquid as the particles begin to stick together. This is how clouds and fog are formed. This is also why windows become covered in droplets of water on a cold day. The cold air cools the window so that the water vapour in the air that touches the window turns into water liquid.

This change from gas or vapour into a liquid is called condensation. We often call the liquid on the window 'condensation' but this is not very scientific. We should say that it is droplets of water caused by condensation.

Activity – catch moisture from the air

You will need:

- a bottle with a screw lid
- crushed ice
- very cold water
- a towel.

Fill the bottle with as much crushed ice as possible and then top it up with the very cold water. Screw the cap on tightly.

Use the towel to dry the outside of the bottle.

Put the bottle somewhere in the classroom and leave it for a while.

After a short time you will see that droplets of water appear on the outside of the bottle. When the water vapour in the air meets the cold surface of the bottle, it condenses into liquid water.

Water vapour from the air condenses when it meets a cold glass

Not all materials will change state when they are heated. Sometimes heating a material causes a different type of change to take place. This type of change makes new materials and is not reversible. For example when you bake cookie mixture or bread dough in the oven, you cannot change it back afterwards. These changes are called chemical changes and you will learn more about them in Year 5.

Did you know?

The Namibian desert is one of the driest places on Earth. It almost never rains but each night, when the air cools down, water vapour in the air from the nearby ocean condenses into droplets, making a fog.

There is a type of beetle called the Namibian fog-basking beetle. It spends the hot, dry days hunting for food, burying itself in the sand when it gets too hot. During the cool nights, it comes out and does a sort of handstand. Droplets of water from the night fog condense onto its back and roll down its shell into its mouth so it can have a drink.

Exercise 5.3a

1 What name do we give the change from liquid to solid?

2 At what temperature does water turn from a liquid to a solid?

3 If we warm some water it will begin to evaporate. What does this mean?

4 What is the boiling point of water?

5 Explain in your own words why droplets of water form on a cold window.

Exercise 5.3b

Use the following words to help you to fill in the gaps in these sentences.

condense evaporates freezing liquid solid vapour

1 The change from liquid to solid is called _____.

2 When we heat liquid water, it _____ to form water _____.

3 When chocolate melts, it turns from a _____ to a _____ .

4 Water vapour in the air will _____ and form droplets of liquid on a cold window.

Exercise 5.3c: extension

Suggest, using your knowledge of the movement and arrangement of particles, why puddles dry up more quickly on a sunny, windy day than on a cool, still day.

 ## Activity – changes of state

You have learnt that different materials change state at different temperatures.

Find the melting points of the following materials:

- iron
- butter
- candle wax
- copper
- oxygen.

Find the boiling points of the following materials:

- oxygen
- alcohol.

Challenge

Look carefully at the boiling points and melting points you have found. The normal human body temperature is 37 °C. Room temperature is about 20 °C.

Can you use this information to explain the following statements?

- Butter is solid at room temperature but melts quickly in your hand.
- Oxygen in the air in this room is a gas.
- It takes a lot of energy to melt iron.

➲ The water cycle

You have learnt that water is always evaporating into the air. We also know that, from time to time, it falls out of the air in the form of rain. The water is changing from liquid to vapour and back again all the time in a constant cycle called the **water cycle**.

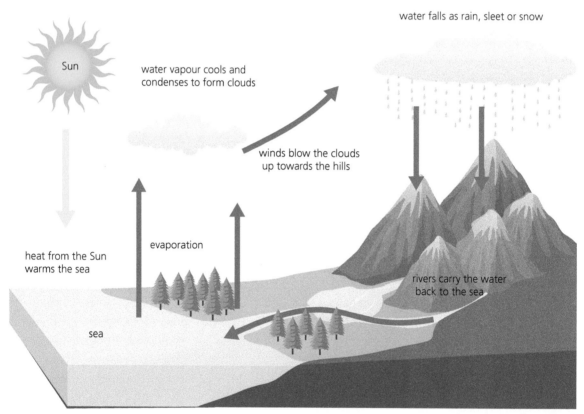

water falls as rain, sleet or snow

Sun

water vapour cools and condenses to form clouds

winds blow the clouds up towards the hills

heat from the Sun warms the sea

evaporation

rivers carry the water back to the sea

sea

■ The water cycle

The heat from the Sun, and the wind make water evaporate from the sea into the air as water vapour. As it rises and cools, the water vapour condenses to form little droplets of liquid, making clouds.

When the clouds are blown inland, they rise up over hills and mountains and become even cooler. The droplets of water get bigger and bigger and finally fall as rain. If it is really cold, the droplets might freeze and fall as snow or hail.

The rain falls onto the Earth and runs into rivers. The rivers carry the water back to the sea and so the cycle begins again.

Activity – make your own rain

You will need:

- a large glass jar
- water
- cling film
- a rubber band
- two or three cubes of ice sealed in a plastic bag.

Put some water in the bottom of the jar.

Put cling film over the mouth of the jar and fasten it with the rubber band.

Make sure that the ice is sealed in the bag so that nothing can leak out and then place the bag carefully on top of the cling film.

If possible, put the jar in a warm place, for example on a sunny windowsill, near a radiator or under a lamp.

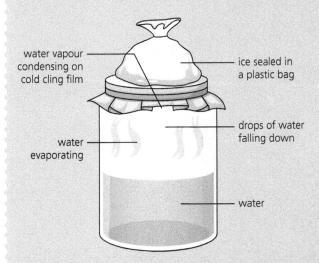

water vapour condensing on cold cling film

ice sealed in a plastic bag

drops of water falling down

water evaporating

water

Watch carefully and you should be able to spot droplets of water forming on the cling film where the ice has cooled it. When the droplets get big enough they will drop down forming 'rain' in your jam jar. You have made a little model of the water cycle.

Sound and hearing

Sounds are all around us. If you stop and listen for a few moments, you will hear lots of different sounds. Sometimes we don't really notice them because we hear them so often. Sometimes they can be a bit of a shock.

Look at this picture. How many different sources of sound can you spot?

➲ How are sounds made?

All the sounds you spotted in the picture are different. Some are loud; some are quiet. Some are high and squeaky; some are low and rumbling. Every one of them has one thing in common. The object that is making the sound is **vibrating**.

The word 'vibrate' means to move backwards and forwards very quickly. The differences between the sounds are caused by differences in these movements, called **vibrations**.

Sometimes you can see the vibrations. If you stretch an elastic band between your fingers and pluck it, you can see the vibrations that make the sound.

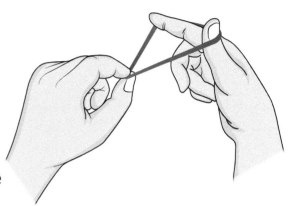

Often you cannot see the vibrations when something is making a sound. If you blow a recorder, you can hear the sound but you cannot see anything vibrating. Sometimes we can make these invisible vibrations easier to see.

 Activity – vibrations

Tuning fork

You will need:

- a tuning fork
- a shallow dish of water.

Strike the tuning fork gently on the edge of the table. Look at it carefully. You may be able to spot that the prongs are vibrating quite fast.

Strike the tuning fork again and then dip the tips of the prongs into the dish of water.

Why does the water move and splash about? Discuss your ideas with your partner or group.

Drum

You will need:

- a drum (or a biscuit tin with cling film tightly stretched over the opening and secured with an elastic band)
- a drum stick
- salt or uncooked rice.

Tap the skin of the drum gently with the drumstick. Look at it carefully. Can you see the skin vibrating?

Now sprinkle a little salt or a few grains of rice onto the drum skin. Tap the drum again like you did before.

Can you explain what happens to the salt or rice? Discuss your ideas with your partner or group.

➔ Making music

An object makes a sound when it vibrates. However, different objects make different sounds. Sometimes one object can make different sounds. Musical instruments can often make lots of different sounds, depending on how you play them. People change the sounds so that they all go together to make music.

Activity – musical instruments

You will need:

- a variety of different musical instruments.

Make a sound with each of the different instruments. Watch carefully. Can you see anything vibrating?

Discuss your observations with your partner or group. If you cannot see the vibrations, can you predict what might be vibrating? Compare your ideas with the rest of the class.

Next, take each instrument in turn and see how many different ways you can find to change the sound you make.

Musical instruments are grouped together according to the way in which they make sounds. You may have learnt about this in your music lessons.

Stringed instruments, such as a violin, cello or guitar, make a sound when the strings vibrate. We make this happen by drawing a bow across the strings or by plucking them.

Wind instruments make sounds when the air inside them vibrates. Players of **brass instruments**, such as a trumpet or French horn, make the air vibrate by vibrating their own lips against the mouthpiece.

Woodwind instruments, such as an oboe or clarinet, often have a reed, which is a thin piece of woody material. This vibrates when air is blown past it.

Percussion instruments vibrate when they are hit or shaken. For example a cymbal or triangle will vibrate when it is hit with a stick.

■ People can play different instruments together in a band or orchestra to make music

Use the following words to help you to fill in the gaps in these sentences. Some words may need to be used more than once.

air   percussion   strings   vibrate   vibrations

1 Sounds are made when objects _____.

2 Sometimes the _____ can be seen, but often they cannot.

3 Instruments such as violins and guitars make a sound when the _____ vibrate.

4 Instruments that make sounds when they are hit or shaken are called _____ instruments.

5 Wind instruments make a sound when the _____ inside them is made to _____.

⊙ Changing sounds

There are two main ways in which we change a sound. We can make the sound louder or quieter. We can also make it higher (more squeaky) or lower (more growling).

Loud and quiet

The loudness or quietness of a sound is called its **volume**. You probably found out how to make the sounds louder and quieter when you were investigating the different musical instruments in the activity above. Did you spot any patterns in your investigation? We can usually make a very quiet sound and then make it gradually louder and louder. How do we do this?

⊙ Volume and vibrations

You will need:

- a drum

- other instruments that make sounds in different ways.

Start with the drum. Make a really quiet sound with it. Now make the sound a little louder, then louder still.

Discuss with your partner or group how you make the sound gradually louder.

Scientists like to describe patterns in a special way. In this investigation they might say:

The harder I hit the drum, the louder the sound it made.

Now do the same with the other instruments. Make a quiet sound and then make it louder and louder. What are you changing to make it louder?

Work with your partner or group to make up a pattern sentence, like the one for the drum above, for each instrument.

When we change the volume of a sound, we change the size of the vibrations. *Loud* sounds are made by *large* vibrations. When a drum is hit really hard, it makes really big vibrations. This makes a loud sound.

Quiet sounds are made by *small* vibrations. A gentle tap on the drum makes tiny vibrations. The sound will be quiet.

High and low

When we play a tune on a musical instrument we change how high or low each note is. This is called

A loud sound is made if you bang a drum hard

the **pitch** of the note. When you investigated the instruments in the activity above, you probably found that you could change the pitch of the sound that most of the instruments made.

To change the pitch of a note, we have to change the speed of the vibrations.

⊃ A vibrating ruler

You will need:

- a ruler

- a table or bench.

Place the ruler on the edge of the table or bench so that about half of it is sticking out.

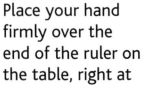

Place your hand firmly over the end of the ruler on the table, right at the edge of the table, and gently flick the other end. Do not flick it too hard or it might snap.

Listen carefully to what you can hear. Watch the vibrations.

Now move the ruler so that only about a quarter of it is sticking out. Hold it down on to the table as before and flick the end.

What do you notice about the sound? Is it higher or lower than before?

What do you notice about the vibrations? Are they quicker or slower than before?

Now move the ruler so that about three quarters of it is sticking out and flick it as before.

Is the sound higher or lower? Are the vibrations quicker or slower?

Discuss your observations with your partner or group. Working together, see if you can make a pattern sentence like we did before.

6   Sound and hearing

When a sound is low-pitched, the vibrations are slow. A high-pitched sound is made by fast vibrations. Your pattern sentence might have been:

When the sound gets lower, the vibrations are slower.

We change the pitch of the note made by musical instruments in different ways.

Strings

Stringed instruments usually have several different strings on them. If you look carefully at a violin or a guitar, you will see that these strings are not all the same thickness. What difference do you think this would make? Violin players will check and change the sound made by each string before a concert so that each violin sounds the same. How do they do this?

➜ Tissue box guitar

You will need:

- an empty tissue box
- two or three elastic bands, the same length but different thicknesses.

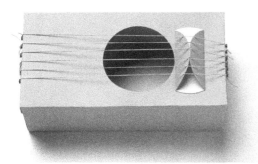

■ A simple guitar can be made from a tissue box or shoe box

Stretch the elastic bands round the tissue box. Try to make sure that they are evenly stretched all round. Your teacher may need to help you with this.

Pluck each elastic band gently with your finger. Which one makes the lowest note – the thinnest one or the thickest one?

Working Scientifically

Can you complete this pattern sentence?

The thicker the elastic band the _____ the note.

Try other ways to change the pitch of the sound. What happens if you shorten the vibrating part of the elastic band by pressing down on it with your finger? What difference does it make if you stretch the elastic bands more tightly? See if you can make up any pattern sentences to describe your findings.

Now see if you can play a simple tune on your tissue box guitar.

Changing the pitch of the note made by a stringed instrument can be done in several ways.

A violinist can play different notes by pulling the bow over the different strings. Thick strings vibrate more slowly than thin ones. The *thicker* the string, the *lower* the note the string makes when it vibrates.

To tune the strings before a concert, the violinist will tighten or loosen the strings, using the pegs at the end of the neck of the violin. Loose strings vibrate more slowly than tight ones. The *looser* the string, the *lower* the note the string will make.

To make lots of different notes with each string the violinist will press down onto the string in different places. This makes the vibrating part of the string shorter or longer. Long strings vibrate more slowly than short ones. The *longer* the string, the *lower* the note the string makes.

Big instruments usually make lower sounds than small ones. Think about a cello. It is larger than a violin. It has thicker and longer strings so it makes

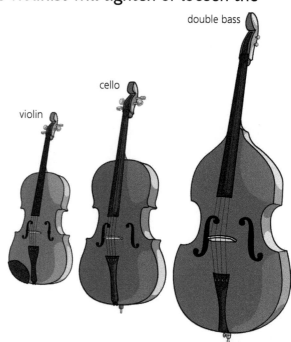

double bass

cello

violin

■ Larger instruments make lower pitched sounds than smaller ones

a range of lower pitched sounds. A double bass is even bigger so it makes even lower pitched sounds.

 Did you know?

The sitar is an Indian stringed instrument. It has up to 20 strings. Each one can be tuned with its own peg. The player will often tune them differently for each piece of music.

■ An Indian sitar has up to 20 strings

Wind

If you play the recorder, clarinet or flute you will know that we change the pitch on wind instruments by closing and opening the holes with our fingers. This changes the length of the column of air that is vibrating. If you cover all the holes, the air all the way down the instrument is vibrating and the note will be low. If some of the holes are open, the air can escape more quickly. The vibrating column of air is shorter and the sound is higher pitched.

➲ Musical straws

You will need:

- straws

- scissors.

Take a straw and flatten one end by pressing it between your fingers.

Use the scissors to cut the end to a point.

Put the pointed end into your mouth; close your lips gently around it and blow hard. If you do not get a sound, change the length of the straw that is inside your mouth a little or maybe how tightly you close your lips against it.

Once you have managed to get a sound from your musical straw, listen to it carefully.

Now take another straw and cut it a bit shorter than the last one. Do you think this will make a higher or lower note than the last one? Make the pointed end as before and blow it to test your prediction.

Make several musical straws, all different lengths.

Which of these pattern sentences is correct?

The shorter the straw, the higher the note.

or

The longer the straw, the higher the note.

Can you work with your partner or group to play a tune on your musical straws?

Did you know?

There is a wind instrument, known as the panpipes or pan flute, which works rather like your musical straws. It has several tubes of different lengths, all joined together. To play the instrument, the player blows across the top of each pipe to make the note.

Panpipes have been played in South America for centuries and are still popular today. Can you guess why they are called panpipes? If not, see if you can find out.

■ Panpipes are popular in South America

⬆ Bottle music

You will need:

- a few glass bottles, all the same size, preferably clear glass
- water in a jug
- a funnel
- a pencil or stick.

Carefully pour a different depth of water into each bottle. You may need to use a funnel if the necks of the bottles are small. Remember to wipe up any spilt water immediately, especially if it is on the floor.

Tap the shoulder of each bottle gently with the pencil or stick to make the bottle vibrate. Listen carefully to the different sounds.

Arrange the bottles in order, with the highest pitched one at one end and the lowest pitched one at the other. Look at the depth of water in the bottles. Can you see a pattern?

Try to play a tune with your bottles. If a note does not sound right you can change it by adding or removing some of the water.

There is another way to make sounds with your bottles. Pick up a bottle and blow gently across the top to make a note. This makes the air in the bottle vibrate.

Try all your bottles. What differences do you notice between the sounds made by blowing and the ones made by tapping the bottle?

Percussion

Percussion instruments are played by hitting or shaking them. Many of them can only make sounds of one pitch but some can make different notes.

A xylophone is played by hitting bars made from wood. Each bar is a different length. Which notes will make a higher pitched sound – the long bars or the short bars? If your school has a xylophone, you could try it out to test your prediction.

Did you know?

People in many different countries all round the world make music with groups of percussion instruments.

In Indonesia, you might find a gamelan orchestra playing gongs, drums and other percussion instruments of different sizes.

In the West Indies you might see a steel band playing special drums made from old metal oil drums.

In many African countries people play drums of different sizes. Great big drums make a deep booming sound. Smaller drums are used to make higher pattering sounds. When lots of them play together it can be really exciting.

Exercise 6.2a

1 Make a list of musical instruments. Sort them into groups according to what vibrates to make the sounds.

2 What word is used to describe the loudness of a sound?

3 Fred's teacher says he is playing his recorder too loudly. What should he do to make the sound quieter?

4 How would you make a low note on a recorder?

5 Give three ways in which the pitch of a note made by a stringed instrument could be made higher.

6 Why does a drum make a louder sound when it is hit harder?

Exercise 6.2b

1 Each of these sentences describes a change in the way an instrument is played. For each one, say whether the sound will be *higher*, *lower*, *louder* or *quieter*.

(a) Play a thicker string on a guitar.

(b) Hit a drum more gently.

(c) Tighten the string on a violin.

(d) Make the vibrating part of a string shorter.

(e) Blow a clarinet harder.

2 What word do we use to describe:

(a) how loud or quiet a sound is

(b) how high or low a sound is?

Exercise 6.2c: extension

Imagine you have found an old guitar in the attic. The strings are loose and when you pluck them they make a horrid sound. Describe how you would adjust the strings to make a nicer sound.

Try to find out what musical note (pitch) each string on a guitar should be tuned to. Can you suggest how you might make sure that the notes are properly tuned?

⊙ Sounds can travel

Sounds are made when an object vibrates. We know that sounds must be able to travel because we can hear them from some distance away.

Listen to your partner speaking to you. The sound comes from your partner's mouth and to your ears so you can hear it. What has the sound travelled through?

The only thing between your ears and your partner's mouth is the air. Sounds can travel through air. The vibrations made by your partner make the air vibrate. This passes the vibrations to your ears.

Air is a gas. Do you think that sounds can also travel through liquids and solids? Discuss your ideas with your partner or group. Have you had any experience of sounds travelling through liquids or solids?

Activity – travelling sounds

You will need:

- a partner

- a table.

Stand at the opposite end of the table to your partner.

Use just one finger to scratch gently on the table. Ask your partner if they can hear the scratching noise.

Now swap over and ask your partner to scratch the table gently. How well can you hear the sound?

Now ask your partner to put one ear against the table and make the scratching noise again. Then swap over so that you put your ear on the table and your partner scratches.

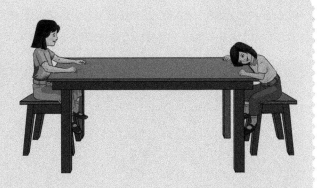

What do you notice about how well you can hear the sound through the table? Was it easier to hear when the sound had to travel through the air to your ear or when it travelled through the solid table?

Sounds travel quite well through air but they travel much better through hard, solid things, like the table. Vibrations can travel a long way through the ground or through hard materials, like steel.

Did you know?

A bird searching for its food will detect a worm by feeling the vibrations of the worm scraping through the soil. The bird feels the vibrations through its feet as well as hearing the sounds of the moving prey with its ears.

Did you know?

The aye-aye is one of the endangered lemurs of Madagascar (see Chapter 2). It has a very long middle finger on each of its front feet. It uses this finger like a drumstick. It taps on the bark of a tree and the vibrations help it to discover if there are insect larvae hidden underneath. If there are, it then uses its strong incisor teeth to bite through the wood to reach the food.

Aye-ayes also eat coconuts and will often tap on coconuts to find out how much liquid there is inside before choosing which one to gnaw.

■ The aye-aye uses its long finger like a drumstick

➜ Can you hear me?

You will need:

- two paper or plastic cups
- a nail
- a piece of modelling clay
- a piece of string, 4–5 metres long.

Push the modelling clay gently into the bottom of one of the cups.

Push the nail through the bottom of the cup and into the modelling clay to make a neat hole in the bottom of the cup.

Take out the modelling clay.

Thread one end of the string through the hole in the cup and tie a big knot in the end of the string inside the cup. This will stop the string from slipping out of the hole again.

Now make a hole in the bottom of the other cup in the same way.

Thread the other end of the string through the hole and tie a knot in the end inside the cup. Now you have made a string telephone.

Ask your partner to hold one of the cups and hold the other one yourself. Move apart so that the string is pulled tight. Do not pull too hard or you will pull the string out of the cups.

Hold your cup to your ear and ask your partner to speak quietly into their cup. Can you hear what is being said?

Now swap over and let your partner hear you speaking. Make sure that you always keep the string tight or your telephone will not work.

Now discuss with your partner what changes could be made to your string telephone. What difference do you think these changes would make?

Plan an investigation. You might decide to investigate if the thickness of the string makes a difference. Perhaps you could find out if the size of the cups makes a difference.

Predict what you think you will discover in your investigation and write your prediction in your notebook.

How will you make sure that your investigation is a fair test?

Ask your teacher for the materials you need and carry out your investigation.

When you have finished, tell the class about what you have done and what you have found out. Remember to try to explain your results using what you have learnt about sounds in this chapter.

Sounds can travel through liquids too. You have probably noticed that you can hear sounds when you are under the water in a swimming pool.

Many sea animals, such as whales and dolphins, make sounds to communicate with each other. These sounds can travel a long way through the seawater.

■ Sea creatures such as whales and dolphins make sounds that travel well through the water

Did you know?

Sounds travel about four times quicker through water than through the air.

Sounds travel through steel about 15 times quicker than through air.

Here on Earth there are almost always sounds around us: the noise of the traffic, the sound of voices, the rustle of leaves or the wind in the grass. It is very difficult to find a place that is without sound.

Vibrations from sounds need something to travel through. We have seen that they can travel through gases, such as the air. They can also travel through liquids, such as water, and solids, such as the wood in the table.

A place that is completely empty, where there are no solids, liquids or gases, is called a **vacuum**. It would be completely silent in a vacuum because there is nothing there to carry the vibrations.

On Earth, there is air all around us and it is quite hard to make a true vacuum. In space, however, there is no air. Sounds cannot travel through space because it is like a vacuum.

→ Sounds and distance

What do you think happens to the volume of a sound as you get further away from the sound source? Think about how well you can hear someone at the other side of the playground or at the far end of the football field.

→ Spreading the sound

You will need:

- a partner

- a big space, such as the playground or games field.

Ask your partner to stand at one end of the space.

Stand next to your partner and ask them to say something in a normal speaking voice.

Walk a few metres away from your partner, shut your eyes and then ask them to say something different at the same volume as before. Can you hear clearly what was said?

Move a few metres further away and do the same thing again. Then move another few metres away and so on. Your partner should say something different each time but always at the same volume.

How does the distance from your partner affect how well you can hear what is being said?

Now swap over so that your partner can hear what happens.

Work with your partner to make up a pattern sentence to describe what happens as you move further away.

 Go further

We can measure the volume of sounds using a sound meter or sound sensor attached to a datalogger. It is also possible to use a sound-measurement app on a tablet or phone.

The volume of sounds is measured in units called **decibels**.

You will need to find a sound source that always makes a sound of the same volume. This might be an electric bell or buzzer. You might be able to use a computer or mobile device with a sound app.

You will also need a tape measure or metre ruler.

Make a table to record your results.

Distance from source, in cm (or m)	Volume of sound, in decibels

Ask everyone around you to be completely silent when you take each of your volume measurements.

Place the sound sensor very close to the sound source and measure the volume with the sound meter or datalogger. Record this in the table.

Move a short distance away from the sound source. Your teacher will tell you how far to move. Use the tape measure or metre ruler to measure the distance.

Measure the volume of the sound again and write it in the table.

Move further away from the source, and measure the distance and the volume again.

Repeat this a few more times, making sure that you always use equal steps to move further away from the sound source.

Look at your measurements. Can you see a pattern in the results? Does this pattern match the pattern sentence you made before?

Did you know?

The crash of thunder and the flash of lightning happen at the same time. However, light travels much, much faster through the air than sound. The light from the lightning reaches our eyes before the sound of the thunder reaches our ears.

The further away you are from the place where the thunderstorm is, the longer the time between when you see the flash and when you hear the crash.

In a thunderstorm the light reaches our eyes some time before the sound reaches our ears

➲ How we hear sounds

When we hear a sound, the vibrations in the air enter our ears. Inside the ear is a small piece of thin skin, stretched tight like the skin of a drum. This is called the **eardrum**.

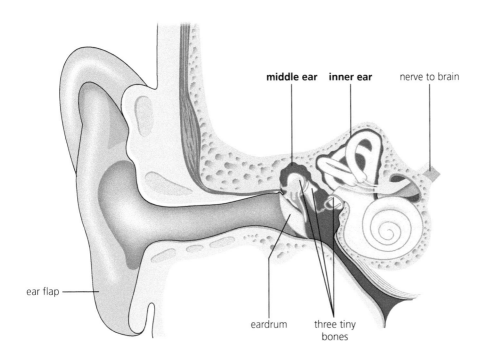

The vibrations entering the ear make the eardrum vibrate. The vibrating eardrum knocks against the three tiny bones in the middle part of the ear and makes them vibrate. These then pass the vibrations into the inner part of the ear.

In the inner ear the nerves change the vibrations into tiny electrical signals. These are then sent to the brain.

Did you know?

We have two ears. Other animals also have two ears – if they have any at all. There must be a good reason why two ears are better than one.

The reason we have two ears is because it helps us to find out where a sound is coming from. The vibrations from the sound may reach the two ears at slightly different times. This is because one ear is further away from the sound than the other. The brain uses this to work out where the sound came from.

The outer part of the ear, often called the ear flap, channels the vibrations into the ear so that we hear the sound as clearly as possible. If you look at a fox hunting you may see that it moves its ears around to help it to pinpoint the exact place where its prey can be found.

■ The fox uses its excellent hearing to help it work out where to find its prey

Changes in hearing

Young children generally have very good hearing. They can hear everything from quite high-pitched sounds to quite low-pitched sounds. This is called their audible range.

Over time, the tiny nerve ends in the inner ear get a bit damaged and do not work as well. Our audible range gets smaller. Gradually we become less able to hear high-pitched sounds as we get older. You might be able to hear the very high-pitched squeak of a bat flying around in your garden. The adults you know are unlikely to be able to hear this sound.

Our ears can become damaged more quickly if we hear lots of loud sounds. If you go to a party where there is loud music or if you listen to loud music through headphones, your ears will be damaged. This damage could reduce your audible range or even make you deaf.

This deafness may wear off after a while but it could be permanent. Remember that very loud sounds are made by very large vibrations. If a sudden burst of very large vibrations meets the eardrum, the soft skin can vibrate so much that it tears. It is important to look after your ears by protecting them from loud sounds.

People who work in noisy jobs wear ear-protectors so that the sounds will not damage their ears. These are made using a material that stops most of the sound from passing through.

⊙ Make ear protectors

You will need:

- two shallow cups or dishes large enough to cover your ears
- a variety of different materials
- a sound source that always makes a sound with the same volume.

Look at the materials you have been given. Which one do you think might make the best ear protector? Discuss with your partner or group why you think this will be the best. Which do you think will be the worst? Why?

Choose one of the materials and pack some of it into the two cups. Use the same material in both cups. Hold one of the cups over each ear and listen to the sound from the sound source. Does the material block some of the sound?

Discuss with your partner or group how you will carry out a fair test to compare the materials you have been given. What is the one thing you will change? What must you keep the same each time? How will you tell which one is the best?

When you have tested all the materials, discuss your results with your partner or group. Which material was the best? Did you all agree? Was the predication you made at the start correct? Can you explain why this material was the best?

When scientists do investigations they usually think about how good their work has been. How well do you think you planned your investigation? How easy was it for you to decide which material was the best? Can you think of anything you might have done to make it better?

Scientific investigations are done to answer questions. Often the results lead to even more questions. See if you can think of any more interesting questions about materials that stop sound from passing through them. How might you find the answers to your questions? Could you do more experiments or will you need to look up the answers somewhere?

Evelyn Glennie is a famous percussion player. She has been completely deaf since she was 12 years old. She does not wear shoes when she performs because she feels the vibrations of the music through her feet and other parts of her body. She can detect the pitch of a note by where on her body she feels the vibrations.

She plays many percussion instruments, some of which you might recognise. She also plays many other instruments from all round the world. Sometimes she makes her own, using bits and pieces she finds, such as scrap metal, flower pots and kitchen utensils. She might use up to 60 different instruments in one concert. She is a very gifted musician who plays very sensitively even though she is deaf.

Exercise 6.3a

1 Does sound travel quickest through solids, liquids or gases?

2 You can hear a dog barking outside the window. What has the sound travelled through to reach your ears?

3 Why is it easy to hear someone talking when you are standing next to them but harder when they are further away?

4 What is a vacuum?

5 Why do sounds not travel through outer space?

6 Explain in your own words how our ears allow us to hear sounds.

7 Some people use very high-pitched whistles to call their dogs. Children can hear the sound but adults cannot. Explain why this is the case.

8 Explain why loud sounds can be harmful to your ears.

Exercise 6.3b

Use the following words to help you to fill in the gaps in these sentences. Some words may need to be used more than once.

brain   deaf   eardrum   gases   high-pitched   liquids   quieter vacuum   vibrations

1 Sounds create _____ in the air.

2 Sounds can travel through solids, _____ and _____ but not through a _____.

3 We hear a sound when it enters our ears and makes the _____ vibrate.

4 Nerves in the inner ear feel the vibrations and send a message to the _____.

5 Loud sounds can damage the _____. This could make you

_____.

6 Sounds become _____ as you travel further away from the sound source.

7 Children can hear _____ sounds more easily than adults can.

Exercise 6.3c: extension

Bats make high-pitched squeaking noises as they fly. Find out how this helps them to fly around safely at night and catch their prey. See if you can also find out how people studying bats can use these squeaks to work out which type of bat is present.

Electricity

Electricity is a very useful kind of energy. We use electricity every day. Our lives would be extremely different if we suddenly had to do without it. Have you ever had a power cut at home or at school? Think about what it was like. What could you *not* do while the electricity was off?

We use electricity without really thinking about it. It is very easy to plug something into the mains socket and turn it on. Everyday objects, such as the fridge, the washing machine and the television need to be plugged into an electricity supply to work.

Sometimes we need to carry things around with us. For example a mobile phone and a torch would not be much use if they had to be plugged in all the time. They still use electricity though, so it is a good thing that we can also get a supply of electricity from batteries.

Activity – things that use electricity

Look around the room. Make a list of everything you can see in the room that uses electricity to make it work.

Now think about everything that you have done so far today. Add to your list anything else you have used today that uses electricity to make it work.

Compare your list to the ones made by the rest of your group. Did they think of anything you missed?

It is important that we understand a bit about electricity so that we can use it safely. People knew about the sparks made by electricity for a long time before anyone worked out how to make it useful. We can start our study of electricity by looking at why these sparks happen.

 Go further

⊙ In the beginning

Long ago, people discovered that some materials make sparks when they are rubbed together. You may have seen this sometimes when you have taken off your jumper in the dark. You may have felt a spark as a tiny prickly shock when you touched something. These sparks are usually called **static electricity**. The word 'static' means 'staying still'. Static electricity stays still unless it becomes possible for it to jump from one place to another, making a spark as it goes.

When two surfaces rub together, tiny bits of electricity move from one to the other. When this happens, we say that the surfaces have become **charged** with static electricity. There are two types of static charge: positive (+) and negative (−). These behave a bit like magnets. A positive charge and a negative charge will **attract** each other. Two positive, or two negative charges will **repel** each other. We can use this fact to play some fun games with static electricity.

 Activity – games with static electricity

Fighting balloons

You will need:

- two balloons
- some thread
- a woolly jumper or cloth.

You have probably seen how a balloon that has been rubbed on a jumper will stick to the wall or ceiling. Here's another trick with balloons.

Take two balloons, blow them up and tie threads onto the necks so that they can be hung up. Rub both balloons on a woolly jumper or on a cloth. It is best to rub both balloons with the same material for this trick. This gives them both the same kind of charge.

Hold the threads and let the balloons hang downwards. Try to bring the balloons together. Can you explain why they push each other away?

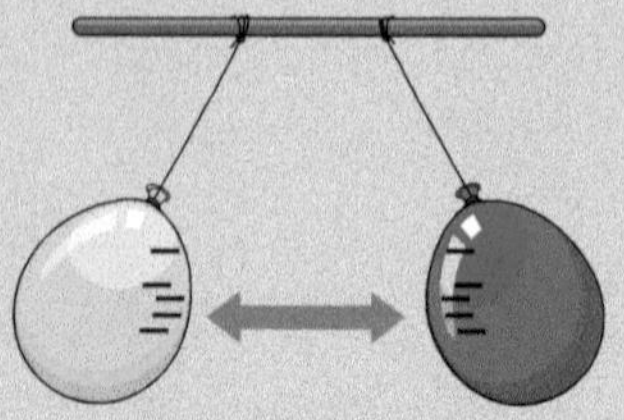

Obedient water

You will need:

- a plastic ruler or comb, or a balloon
- a woolly jumper or cloth
- access to a water tap.

■ Why do the balloons repel each other?

Take a balloon, or a plastic ruler or comb and rub it on the jumper or cloth.

Turn on a cold tap and adjust it so there is a thin stream of water coming out. Hold the rubbed balloon, ruler or comb near the stream of water. What happens?

You will see a bigger effect if you do not put the ruler, comb or balloon too close to the tap.

Wake up sleepy Joe

You will need:

- tissue paper
- scissors
- sticky tape
- a metal pie dish or piece of card covered with kitchen foil
- a balloon, or plastic ruler or comb.

Cut a 'sleepy Joe' out of tissue paper, like the one in the picture. Do not make him too big. He needs to be able to sleep on top of your pie dish or card. Make sure that Joe has big feet!

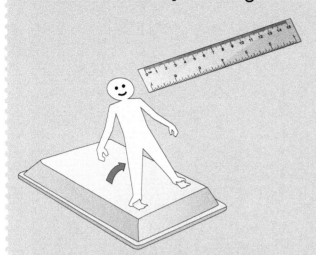

■ How can you wake up sleepy Joe?

Use sticky tape to stick Joe's feet to the metal pie dish or the sheet of card covered in kitchen foil.

Rub your balloon, ruler or comb on the jumper or cloth again and hold it near Joe's head. Can you make Joe stand up?

Hair raising

You will need:

- a balloon

- a woolly jumper or cloth.

This works best with a balloon but you could try it with a plastic ruler or comb.

Rub your balloon on the jumper or cloth to build up a good charge. Hold the balloon close to your partner's head. What happens to your partner's hair?

Let them try with your hair. Is there any difference? If so, can you think of any reason why?

In 1929, an American scientist called Robert Van der Graaf, invented a machine that can make a much bigger static charge than rubbing a balloon. If you put your hand on this machine, called a Van der Graaf generator, it can really make your hair stand up on end! It also makes a big spark if the static electricity jumps away from it, so it must be used carefully.

If you are very lucky, your teacher may be able to show you a Van der Graaf generator like the one in the picture.

■ A Van der Graaf generator makes a much bigger static charge than a balloon

Dancing cereal

You will need:

- a clean, dry plastic bottle
- puffed rice cereal
- a balloon, or a plastic ruler or comb.

Take the clean, dry plastic bottle (it must be very dry). Put a few pieces of puffed rice cereal in the bottle and screw on the cap.

Rub your balloon, ruler or comb on the jumper or cloth again and bring it near the bottle.

What happens to the cereal grains? Try rubbing the bottle with the jumper or cloth to see what happens then. Can you think why this happened?

Exercise 7.1

Use the following words to help you to fill in the gaps in these sentences. Each word may be used once, more than once or not at all.

attract　　**negative**　　**positive**　　**repel**　　**static**

1 Electricity that stays still on a surface is called _____ electricity.

2 There are two types of static charge, called _____ and _____ charges.

3 Two surfaces with the same charge on them will _____ each other.

4 Two surfaces with the opposite charge on them will _____ each other.

➲ Moving electricity

Static electricity is fun to play with but it is not very useful. You cannot make toys, computers or bedside lamps work with static electricity.

To be useful the electricity needs to move. Then it can carry electrical energy to the places where we want to make it do some work. We can do this by putting it into a sort of electrical 'running track' called a circuit.

A circuit needs at least three things in it: a source of electrical energy, something for the electricity to run through and something to work when the energy reaches it. The things that make up a circuit are known as components.

The source of electrical energy could be the mains electricity supply. When we are doing experiments in the classroom we do not use this because it can be very dangerous. You will learn more about this later in this chapter.

Instead, we use batteries because they are much safer. The proper name for what we usually call a battery is a **cell**. This is the name we will use in this chapter.

When we make a circuit, we use wires to join it up. Electricity can run easily through wires so these make the pathway for it to flow around the circuit.

The electrical energy can make things work. It could light a bulb, make a bell or buzzer sound or maybe make a motor turn.

■ Mains electricity is useful but it is safer to use batteries for experiments.

Light the bulb

You will need:

- a cell

- two pieces of wire (if the wire has a plastic coating on it, make sure that there is a bit of bare metal wire at each end)

- a small light bulb.

Your challenge is to find ways of making the bulb light up using the cell and one or two pieces of wire.

When you find an arrangement that works, draw a neat sketch to show how you did it.

How many different ways can you find to make the bulb light? Can you explain to your partner why some arrangements work and some do not?

Working scientifically

Look carefully at the cell. It has two different-looking ends, called **terminals**. The end with a little bump on it is called the positive terminal and it is usually marked with a '+' sign. The flat end is called the negative terminal and this may be marked with a '–' sign.

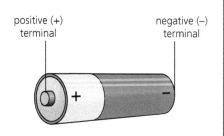
positive (+) terminal   negative (–) terminal

The bulb has two terminals too. One is the little metal button on the bottom. The other one is the screw thread.

Look carefully at the bulb. Your teacher may let you use a hand lens or magnifying glass. Can you see the very thin piece of wire inside the bulb? This is called the **filament**. When the electricity flows through the filament, it gets very hot and glows. One end of the filament is connected to each of the two terminals of the bulb.

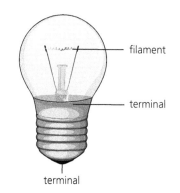
filament

terminal

terminal

To make the bulb light up, the terminals of the bulb and cell must be joined up so that a pathway, or circuit, is made for the electricity to flow round and round. We call this a **complete circuit**.

If there is a gap in the circuit the electricity cannot flow across the gap. This means that no energy reaches the bulb so it cannot light up. It is important that there is a complete pathway from the positive terminal of the cell to one terminal of the bulb. There must also be a pathway from the other terminal of the bulb back to the negative terminal of the cell.

Tiny charged particles pick up energy from the cell, carry it to the bulb and then go back to the cell to pick up some more. They go round and round the circuit, all moving in the same direction, until the circuit is broken.

If we have a simple circuit with just one possible route for the electricity to flow around, we call it a **series circuit**. When the parts of a circuit are joined up in this way we say that they are **in series**.

Exercise 7.2

1 How can you tell the difference between the positive and negative terminals of a cell?

2 Draw a simple picture of a light bulb and label the two terminals.

3 How must a cell, a bulb and some wires be connected to make the bulb light up?

4 Here are some circuits. Will the bulbs light up? For each one say 'Yes' or 'No'.

Circuit		Circuit	
A		D	
B		E	
C		F	

You may have found the last activity a bit fiddly. To make it easier to make circuits we usually put the components into special carriers. These have sockets on them. We can join them together with wires, called **leads**, which have plugs or clips on the end. Your teacher will give you some of these for the next activity.

Make some series circuits

You will need:

- a cell

- some leads

- a light bulb

- a buzzer

- a motor with a small propeller attached.

Start by joining the cell and a bulb together to make a circuit.

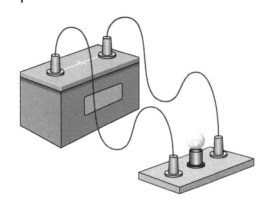

Now unplug the leads from the cell, turn the cell round and plug it back in again. Does this make any difference to the bulb?

Now see if you can make the motor work. Remember that you need to make a complete circuit with just one pathway for the electricity to flow round.

Draw a neat sketch to show how you joined your circuit up.

Now unplug the leads from the cell, turn the cell round and plug it back in again. Does this make any difference to the motor?

Next try to make the buzzer work. Draw a sketch to show how you did it. Does it matter which way round the cell is connected to the buzzer?

The circuits you have made so far have had just one cell and one other component joined together. We can change the number of components in a circuit and this changes the way the circuit works.

More circuits

In this activity you are going to investigate what happens when you change the number of components in a circuit. You need to work carefully and sensibly to try and find a pattern in your observations.

You will need:

- three cells
- three light bulbs
- some leads.

Start by making the circuit with one cell and one bulb again as you did before.

Now discuss with your partner or group what you think might happen if you change the number of bulbs or the number of cells in the circuit.

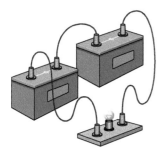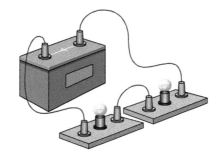

Now carry out the activity to test your predictions. Remember to change one thing at a time. If you do not, you will not be able to tell which thing causes any differences you notice.

When you have carried out your tests, work with your partner or group to write some pattern sentences to describe the patterns you have noticed.

Your first pattern sentence might start:

The more cells we add to a circuit ...

Now work with your partner to see if you can suggest a reason for each of the patterns you have discovered.

If you have time, make a circuit with one cell and a motor. What do you think will change if you add another cell to the circuit? Test your prediction by adding another cell to the circuit. Was your prediction right?

When the bulb lights up, it uses energy. The more energy it has, the brighter it will be. When you add another cell to the circuit, there is more energy for the bulb so it glows more brightly.

If there are several components in a series circuit, they have to share the energy between them. If you add another bulb to your original circuit, there are two bulbs sharing one cell's energy so they will not be so bright.

You will learn more about these patterns in Year 6.

➲ Conductors and insulators

When we make a circuit we want the electricity to run through the different components. Some materials allow electricity to run through them very easily. We call these materials **electrical conductors**. Others act like a barrier and stop the electricity from moving. We call these materials **electrical insulators**.

Investigate electrical conductors and insulators

You will need:

- a cell
- a light bulb
- three leads
- some materials to test, such as copper, cardboard, plastic, steel, pencil lead, iron, fabric, wood, rubber
- two crocodile clips (optional).

Connect one bulb and one cell in series, as you did before.

Working scientifically

Now make a break in the circuit and add another lead as shown in the picture on the right. If you have some crocodile clips you can add them to the loose ends of the leads.

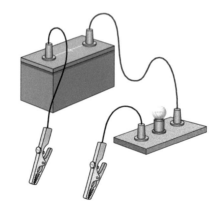

You now have a circuit with a break in it so the bulb is not lit. Touch the ends of the leads or the crocodile clips together to check that your bulb still lights up.

You are now going to test each of your materials to see if it can make a bridge between the loose ends of the leads and mend the gap in the circuit.

To test your materials, put the end of one of the leads onto the material. If you are using crocodile clips you can clip one on to the material.

Now touch or clip the end of the other lead onto the material, making sure that it is a little way away from the other one.

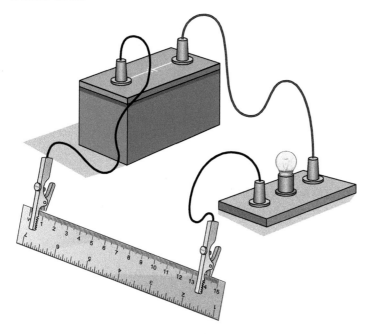

Look to see if the bulb lights up. If it does, then the material has made a bridge that the electricity can flow through and the circuit is complete. The material is an electrical conductor.

If the bulb does not light up, then the material is an electrical insulator. The electricity cannot flow through it so it has not mended the gap in the circuit.

Test the other materials in the same way. Record your results in a table like the one below.

Material	Does the bulb light up?	Electrical conductor or electrical insulator?

Can you see a pattern in your results? Look at all the materials that are electrical conductors. Can you see anything that they all have in common? Are there any exceptions to this rule?

In the activity above, you found that some materials are good electrical conductors and some are electrical insulators. You should have spotted that all metals are good electrical conductors.

Materials that are not metals (known as non-metals) are usually electrical insulators. There is one non-metal that does conduct electricity well. This is the material that is used to make the leads in most writing pencils. It is called graphite.

If we know which materials are electrical conductors and which are electrical insulators, we can use this to help us to choose materials to use to make things. For example, the wires that carry the electricity around your house are made from a metal called copper because copper is a good electrical conductor.

We can also use this information to keep ourselves safe. For example, the plugs on electrical appliances are covered in plastic. Plastic is a good electrical insulator so the plastic plug cover helps to make sure that the electricity does not flow through into you and hurt you. You will learn more about safety later in the chapter.

Sometimes a substance can behave in very different ways depending on how the tiny particles that make it are arranged. Graphite is a soft black material made from a substance called carbon. If carbon particles are arranged in a different way they can make diamond. Diamond is one of the hardest materials known and does not conduct electricity. There are lots of other forms of carbon. Each one has very different properties.

➡ Switches

Make a series circuit with one cell and one bulb as before but this time add a switch. What does the switch do to the circuit?

A switch is a device that opens and closes a gap in the circuit. When the gap is open, the electricity cannot flow so the circuit stops working. When the switch is closed, the electricity can flow through the closed switch. The components in the circuit work because there is no longer a gap in the circuit.

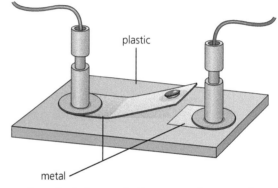

■ A switch is made from a mixture of electrical conductors and insulators.

Think about how your switch opens and closes. Look at the materials that were used to make the switch.

Discuss with your partner or group which parts of the switch are made from electrical conductors and which are made from electrical insulators. Can you explain how the switch works?

Working scientifically

Make a switch

Your teacher will give you some materials to work with. Some will be electrical insulators and some will be electrical conductors.

Use the materials to make a switch that you can connect into your circuit and use to switch the bulb on and off.

Draw a neat picture of your switch and label it to show where the electrical conductors and insulators are.

Show your switch to the rest of your class. Describe how you made it and explain why you chose to use each of the different materials.

Listen to the rest of the class talk about their switches. Which ones do you think were the most successful? Why were these ones so good? Can you think of any changes that could be made to your switch to make it better? Make a note of these changes on your labelled drawing. Remember to explain why these changes would improve your switch.

 Exercise 7.3a

1 Emma made a circuit with one cell and one bulb. Her bulb was not very bright. What should she do to make it brighter?

2 Hassan wanted some electric lights in his toy garage. He made a circuit with one cell and two bulbs. Explain why his bulbs were not very bright.

3 Describe how you can test to find if a material is an electrical conductor or an insulator.

4 Charlie says only metals are electrical conductors. Becky says some non-metals conduct electricity too. Who is right? Explain your answer.

5 Explain how a switch works.

Exercise 7.3b

Use the following words to help you to fill in the gaps in these sentences. Some words may be used more than once.

brightly   conductors   dimly   graphite   insulators   switch

1 Materials that allow electricity to flow through them are called electrical _____.

2 Materials that stop the flow of electricity are called electrical _____.

3 All metals are good electrical _____.

4 All non-metals are electrical insulators, except _____.

5 If we add more cells to a circuit, a bulb will glow more _____.

6 If we add more bulbs to a circuit, they will all glow more _____.

7 A _____ is a device that opens and closes a gap in a circuit.

➔ Using electricity safely

Electricity is extremely useful. It can also be very dangerous if not used correctly. You will not come to any harm during your experiments using cells and small bulbs because the flow of electricity produced by cells is very small. A small cell produces a very small amount of electrical energy and if it flowed through your body, you would not even notice it.

However, the mains electricity that comes into the wall sockets in your house has a lot more energy. If that goes through your body it could cause you a lot of harm and even kill you, so it is very important to know how to use electricity safely.

Here are six simple rules to follow to help you to stay safe.

1 Never play with mains electricity or fiddle with electrical devices.

2 Never use an electrical device that looks as if it has become damaged.

3 Never push anything other than a properly wired plug into an electric socket.

4 Always make sure that wires are tucked away neatly and do not trail across the floor or furniture.

5 Never plug lots of devices into one socket. Extension bars with a few sockets are safe but you should never plug an extension bar into another one.

6 Always keep electric devices away from water.

Activity – understanding the safety rules

Look at this picture. There are lots of people doing silly things with electricity.

Work with your partner or group to see how many dangers you can spot. Explain why each of them is dangerous.

➜ Keeping us safe

If something goes wrong in an electric circuit, it could kill someone or cause a fire. It is important to do all we can to stop this happening. Here are some ways that we can keep ourselves safe.

Insulation

If electricity from the mains flows through your body, it might kill you. To stop this happening we can cover all the wires, and other parts where electricity flows, with a material that is a good electrical insulator. Usually this is plastic. If the insulation becomes damaged, the electricity could flow out through the gap. This is why you should never use an electrical device that has become damaged.

Electric plugs and fuses

This is the inside of an electric plug.

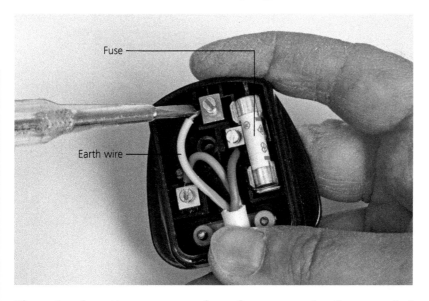

Fuse

Earth wire

Electric plugs have two safety features: the **fuse** and the **earth wire**.

The earth wire is a safe path for electricity to flow away from the plug if something goes wrong.

A fuse has a very thin piece of wire inside. If too much electricity flows into the plug this wire will melt. This makes a break in the circuit and so no more electricity can flow.

Go further

All the electricity comes into your house through a special box called the fuse box. Do you know where this is in your house?

Inside the fuse box are lots of special fuses called circuit breakers. These open a gap in an electrical circuit if there is a problem. There are lots of different electrical circuits in the house. Each one has its own circuit breaker. This means that an electrical problem in one part of the house can be fixed, leaving the rest of the house with everything working as usual.

■ A fuse box being tested by a qualified electrician

Sometimes it is a good idea to add an extra layer of safety by plugging a small circuit breaker into the socket before plugging in a device, such as a lawn mower or electric drill. Can you suggest why a bit of extra safety might be needed for items such as these?

Exercise 7.4a

Use the following words to help you to fill in the gaps in these sentences. Some words may be used more than once.

break   damaged   dangerous   fuse   insulator   plastic

plug   water

1 Electricity can be very _____.

2 You should never use an electrical device that has been _____.

3 Never push anything other than a proper _____ into an electrical socket.

4 Keep electrical devices away from _____.

5 Wires are often covered in _____ because it is a good electrical _____.

6 A _____ in a plug will make a _____ in the circuit if something goes wrong.

Exercise 7.4b: extension

Prepare a short talk about electrical safety to give to the rest of your class.

You could illustrate your talk by preparing a poster or some slides.

Remember to be clear about what the safety rules are and why they are important.

Glossary

Acid rain Rainwater with dissolved acid from factories and power stations.

Adaptation The way in which the features of an animal or plant help it to survive in its habitat.

Amphibian A vertebrate animal with a smooth, damp skin that lays eggs in water.

Antennae The 'stalks' on the heads of some invertebrates, carrying the eyes or other sensing organs.

Archaeologist A scientist who studies human history by digging up human remains and the bits and pieces from their lives that have become buried in the ground.

Atmosphere The layer of gases that surrounds the Earth.

Attract Pull together.

Audible range The range of sounds, from low-pitched to high-pitched, that a person can hear.

Bird A vertebrate animal with feathers and wings on its body.

Boil When bubbles of vapour form in a liquid when it is heated to its boiling point.

Boiling point The temperature at which a liquid boils.

Brass instruments Musical instruments made of metal in which the sound is made by air blown through lips vibrating against the mouthpiece.

Canine A type of tooth, found near the front of the mouth, that is used for tearing food and gripping prey.

Carnivore An animal that only eats other animals.

Cavity (In teeth) a hole in the enamel surface of the tooth.

Cell The proper name for what is usually called a battery.

Characteristic A particular feature of a living thing.

Charged Having a build up of static electricity on its surface.

Circuit A loop containing at least one cell, wires and components, around which electricity can flow.

Community The plants and animals living in a habitat.

Complete circuit A circuit that is properly connected up to give an unbroken pathway for the electricity to flow through.

Component Items that are connected up to make useful circuits, e.g. bulb, motor, buzzer, cell, etc.

Compressible Able to be squeezed (compressed).

Condensation Turning from a gas to a liquid when cooled.

Consumer A living thing that eats (consumes) plant or animal material to get energy. All animals are consumers.

Crown (Of a tooth) the part of the tooth that is outside the gum.

Crustacean An invertebrate animal with an exoskeleton and ten legs.

Decibels Units used to measure the volume of sounds.

Decomposer A living thing that breaks down dead plant or animal material and returns nutrients to the soil.

Dentine The soft inner layer of a tooth.

Dentist An expert in tooth care who will check and mend your teeth.

Digestion The process of breaking down food into tiny particles that can be taken into the blood and carried around the body.

Digestive system The parts of the body that break down and absorb our food.

Eardrum The part of the ear that first detects the vibrations from sounds.

Earth wire A wire in a mains electrical circuit that makes a safe path to carry electricity away if there is a problem.

Electrical conductor A material that allows electricity to flow through it.

Electrical insulator A material that does not allow electricity to flow through it.

Enamel The hard, shiny layer on the outside of a tooth.

Endangered Very few of the species left, so it is in danger of extinction.

Environment The surroundings; the whole of the natural world.

Esophagus The tube running from the mouth to the stomach, also known as the gullet.

Evaporation Turning from a liquid to a gas when heated.

Exoskeleton The hard outside case surrounding the body of some invertebrates, e.g. insects.

Expand Get bigger.

Extinct When there are no more living members of a species.

Fair test A test where one thing is changed and everything else is kept constant so that a fair comparison can be made.

Filament A very thin piece of wire in a lamp or fuse.

Fish A vertebrate animal with a scaly body that lives in water and breathes through gills.

Flowering plant A plant that has flowers and produces seeds for reproduction.

Fluoride A mineral that strengthens tooth enamel.

Food chain A diagram showing how food energy passes from one living thing to another.

Freezing Turning from a liquid to a solid when cooled.

Freezing point The temperature at which a material changes between being a liquid and a solid. This is the same temperature as the melting point.

Frog spawn Frogs' eggs.

Function Job.

Fuse A safety device containing a metal filament that melts if too much electricity flows through it.

Gas One of the three states of matter. Gases can flow and spread out to fill the whole of their container.

Graphite A non-metal that conducts electricity; it is a form of carbon.

Gullet The tube running from the mouth to the stomach, also known as the esophagus.

Habitat The place where an animal or plant lives.

Herbivore An animal that only eats plants.

In series Arranged in a circuit so that there is only one path that the electricity can take around the circuit.

Incisor A type of tooth, found at the front of the mouth, that is used for cutting food.

Insect An invertebrate animal with an exoskeleton and six legs.

Invertebrate An animal without an internal skeleton made from bones.

Key A means of identification of living things based on visible features; may be branching or number.

Large intestine The part of the digestive system where water is removed from the waste materials.

Leads Wires with plugs or clips on the ends used to connect up electrical circuits.

Liquid One of the three states of matter; liquids can flow and take the shape of the bottom of their container.

Mammal A vertebrate animal with fur on its body, which feeds its young with milk.

Melting Turning from a solid to a liquid when heated.

Melting point The temperature at which a material changes between being a solid and a liquid; this is the same temperature as the freezing point.

Metal A shiny material, such as iron, steel or copper, that is a good electrical conductor.

Molar A type of tooth, found at the back of the mouth, that is used for chewing and grinding.

Myriapod An invertebrate animal with an exoskeleton and more than ten legs.

Non-flowering plant A plant that does not have flowers for reproduction.

Non-metal Any material that is not a metal.

Nutrients Substances in our food and drink that are needed to keep the body working properly.

Nutrition Feeding.

Omnivore An animal that eats plants and other animals.

Organ A part of the body that has a particular job to do.

Particles Very tiny pieces of something, so small that they cannot be seen.

Percussion instruments Musical instruments in which the sound is made by hitting or shaking.

Photosynthesis The process used by plants to make their own food.

Pitch (Of a sound) how high or low a sound is.

Plaque A layer on the surface of the tooth made by bacteria.

Pollute To add things to the environment that should not be there.

Pollution Adding things to the environment that should not be there; the things that have been added to the environment that shouldn't be there.

Pre-molar A type of tooth, found near the back of the mouth, that is used for chewing and grinding.

Predator An animal that hunts other animals for food.

Prediction A guess about what will happen based on previous knowledge.

Prey An animal that is hunted by predators for food.

Producer A plant; a living thing that can use light energy from the Sun to make food.

Properties How a material behaves in different conditions.

Pulp The space in the centre of a tooth that contains blood vessels and nerves.

Repel Push apart.

Reproduction Making more members of the species; animals have babies to reproduce; most plants make seeds.

Reptile A vertebrate animal with a dry, scaly skin that lays soft-shelled eggs on land.

Respiration The life process that allows living things to get energy from their food.

Reversible Able to be changed back.

Root (Of a tooth) the part of the tooth that is inside the gum.

Saliva The liquid in the mouth that moistens chewed food to make it easier to swallow.

Scavenger An animal that feeds on dead animals.

Seasons Periods of the year when the weather is different; in Britain, spring, summer, autumn and winter.

Series circuit A circuit in which there is only one possible path for the electricity to follow.

Small intestine The part of the digestive system where digested food moves into the blood.

Solid One of the three states of matter; solids keep their shape when moved from one place to another.

Species A particular type of living thing.

Spider An invertebrate animal with an exoskeleton and eight legs.

State of matter Solid, liquid or gas.

Static electricity Electricity that builds up in one place and stays still.

Steam Tiny droplets of very hot water.

Stomach The part of the digestive system that digests most of the food.

Stringed instruments Musical instruments in which the sound is made by vibrating strings, e.g. violin, guitar.

Tadpole Baby frog or toad.

Terminals (Of a cell or other component) the two ends of the component that must be connected into the circuit.

Top carnivore The last consumer in a food chain, which is not eaten by anything else.

Vacuum An area of space that is completely empty, containing nothing at all, not even air.

Vapour The gas that is made when a liquid evaporates.

Vertebrate An animal that has an internal skeleton/backbone.

Vibrate To move very rapidly to and fro.

Vibrating Moving very rapidly to and fro.

Vibration Small, fast movement to and fro.

Volume (Of an object) the amount of space something takes up.

Volume (Of a sound) how loud or quiet a sound is.

Water cycle How water moves between the oceans and the atmosphere.

Woodwind instruments Musical instruments in which the sound is made by blowing through the instrument, often through a reed.

Index